AF688446

Finding Success in Mathematics: The Operations

Dr. Candace Jones

For permission requests, write to the author at the address below.

Dr.Jones@ExpandedEducationalServices.com

Memphis, Tennessee 38141

This book was edited, formatted, designed, and published by:

UNIQUE PUBLISHING HOUSE, LLC

P.O. Box 750792, Memphis, TN 38175

www.uniquehouse.org

ISBN: 979-8-9877059-2-6

UNIQUE HOUSE

Finding Success in Mathematics: The Operations

Dr. Candace Jones

UNIQUE HOUSE

Introduction

Finding Success in Mathematics: The Operations is a complete foundational program for grades K-8. This is course one in the Expanded Educational Services Success Academy Finding Success in Mathematics program.

Each lesson in Finding Success in Mathematics: The Operations course is timed. It is imperative that students commit their basic facts in all four operations to memory. It matters not that students can get the work done; they must get it done in the specified time. This is the one and only requirement for mastery of this course. It is recommended that you set a timed rule for completion of one second per digit.

Students must successfully master *this* course in order to begin the next ones: Finding Success in The Essentials and Algebra 1. After the successful mastery of these three courses, students are ready to begin College math courses.

The Finding Success in Mathematics program is specially formulated to ensure students are solid in the foundations of mathematics before moving on to more complex topics. The overarching problem in education today is that students do not have a proper foundation and as such are not proficient as they should be and are not able to function at grade level. There is a way to alleviate this problem. Education must return to the basics. For far too long, we in academia have relied on new programs and new concepts which are attached to a calendar. Students have been expected to move forward in curriculums they do not have the proper foundation to master resulting in failure and or low proficiency.

At Expanded Educational Services Success Academy, it is our desire to ensure students receive a proper foundation in mathematics, which is why our courses must be taken in order. It is also why our program spans multiple grade levels. It does not matter if a child is in grade 8, if he or she has not memorized their multiplication facts, they should be enrolled in this first course. Because our concept is the One Room Schoolhouse, there is no shame for older children who have not mastered rudimentary concepts because all students in the room are taught the same material and are expected to master it. It is also why the program caters to large grade level spans. Younger students will feel a deep sense of accomplishment as they master these concepts early.

Our goal for our younger students is that they will matriculate through our entire mathematics program and begin taking online College Math and or Algebra by age 12. It is our sincere desire to see all of our younger students who begin their education with us in the early childhood grades to receive their associate degree before or at the same time as they graduate from High School. Ideally, we'd like to see them complete the Associates and begin working on their bachelor's degree well *before* age 18. It can be done as long as the student is proficient in the basics of all the core subjects.

For many years, education has tried to push students forward in a curriculum they were not academically ready for. They thought extending the school year would help or that pouring more money into failing school districts would be the answer. None of that will take the place of teaching students how to make their own money via their own talents, thus motivating them to excel in the basics of mathematics. It is a formula that we have seen work with our students, and we are now making it available to your students.

Happy Success-FULL school year!

How to Find Success in Mathematics: The Operations

This book is unlike other math textbooks. It was not written with the purpose of encompassing a whole ten-month school year. It was written to give you a proper foundation in mathematics.

Each topic requires you to memorize and solve problems with speed and accuracy. Your success in this course greatly relies on you spending time committing facts to memory and improving your speed. You must be diligent to put the time in. If you do, it will pay off.

Out of all the subjects, math is one of my all-time favorites. It is fun. It is in many cases absolute. I think that's the reason so many people are afraid of math – because it is absolute.

If you have had less than stellar experiences in math so far, I challenge you to retrain your brain to love math again – or perhaps for the first time ever. Remember, your brain believes whatever you tell it. So, tell your brain you love math so that it can focus in and net you comprehension and lovely math achievement. Say this aloud: I love math! Now say it again and truly mean it. You have just set the tone for the school year by telling your brain what you want it to believe. Now when things get hard, don't you tell your brain anything different; instead tell it I can do all things through Christ who gives me strength.

Is there anything too hard for God? Nope. He's got math in the bag and so do you!

~ Candace Jones, Ed.D.

Founder of EES Success Academy

Expanded Educational Services, Inc.

Group Grids & Student Mastery Forms

Grade Level Goals & Mastery Sheet.

This purpose of this sheet is for the instructor and the student to keep up with their mastery throughout this program. It is advisable to repeat this course until the student achieves mastery. Because the program spans several grade levels, it is perfectly fine to procure another book for the next school year if needed.

This graphic organizer is for parents who wish to know approximately what grade level to start their child in the program. Keep in mind, the grade levels are not exact. These are just to give you a general idea of where your child should be. I use the term *should* very loosely. If you would be diligent to follow the Finding Success in Mathematics program, your child will be where they should be when they finish high school. If you follow this program, then by the time your child completes high school, they will in fact be ready to succeed in college, technical school or even the military. The FSI Mathematics program suggests allowing your child to begin taking community college courses in grade nine. Enroll them in college classes according to their strengths. If they are not strong in mathematics, then enroll them in History or Communications first until their strengths and confidence level in mathematics improves. This will also ensure that they are ready to succeed in university before or by the end of high school as well.

Group Grid.

Tear out or make copies of the Group Grids in the next section. At the top, there is a vertical space where you should write the name of each student in your class. The grid is where you keep track of where the whole class is in terms of mastering each objective. This will help you lesson plan for the group more effectively. It will also help you determine how to most effectively help those students who have not mastered objectives. Essentially, the Group Grid helps you see how you can pair together so they can help one another in peer tutoring.

Timed Tests

You might want to remove the Timed Tests from each section of this book before beginning this course. You will want to be sure your students can master each timed session. It is recommended that you set a timed rule for completion of one second per digit for each lesson. Lesson one has 15 problems so the time limit should be 15 seconds. Lesson ten has 15 problems of four digits, so the recommended time should be 120 seconds. You can adjust the time limit as needed for your students. They must master these skills within the time frame you decide. This will ensure they have a great grasp on basic mathematics before moving to the more complex topics of the next course.

Grade Level Goals & Mastery Sheet

Name ___________________ School Year _____

GOALS	Grade Level	ATTEMPT DATE	ATTEMPT DATE	ATTEMPT DATE
1 Digit + 1 Digit Addition & Subtraction	**2-5**			
2 Digit + 1 Digit Addition & Subtraction	**2-5**			
3 Digit + 1 Digit Addition & Subtraction	**2-5**			
4 Digit + 1 Digit Addition & Subtraction	**2-5**			
2 Digit + 2 Digit Addition & Subtraction	**2-5**			
3 Digit + 2 Digit Addition & Subtraction	**3-5**			
4 Digit + 2 Digit Addition & Subtraction	**3-5**			
3 Digit + 3 Digit Addition & Subtraction	**3-5**			
4 Digit + 3 Digit Addition & Subtraction	**3-5**			
2 Digit + 3 Digit Addition & Subtraction	**3-5**			
4 Digit + 4 Digit Addition & Subtraction	**3-5**			
UNIT ONE EXAM				
1 Digit x 1 Digit Multiplication & Division	**4-7**			
2 Digit x 1 Multiplication & Division	**4-7**			
3 Digit x 1 Multiplication & Division	**4-7**			

4 Digit x 1 Multiplication & Division	**4-7**			
2 Digit x 2 Multiplication & Division	**4-7**			
3 Digit x 2 Multiplication & Division	**4-7**			
4 Digit x 2 Multiplication & Division	**4-7**			
3 Digit x 3 Multiplication & Division	**4-7**			
4 Digit x 3 Multiplication & Division	**4-7**			
2 Digit x 3 Multiplication & Division	**4-7**			
4 Digit x 4 Multiplication & Division	**4-7**			
UNIT TWO EXAM				

Grade Level Goals Group Grid

Name __________________ School Year _____

Unit One

GOALS										
1 Digit + 1 Digit Addition & Subtraction										
2 Digit + 1 Digit Addition & Subtraction										
3 Digit + 1 Digit Addition & Subtraction										
4 Digit + 1 Digit Addition & Subtraction										
2 Digit + 2 Digit Addition & Subtraction										
3 Digit + 2 Digit Addition & Subtraction										
4 Digit + 2 Digit Addition & Subtraction										

GOALS										
3 Digit + 3 Digit Addition & Subtraction										
4 Digit + 3 Digit Addition & Subtraction										
2 Digit + 3 Digit Addition & Subtraction										
4 Digit + 4 Digit Addition & Subtraction										

FOR INSTRUCTOR USE ONLY

Grade Level Goals Group Grid

Name ________________ School Year _____

Unit Two

GOALS										
1 Digit x 1 Digit Multiplication & Division										
2 Digit x 1 Multiplication & Division										
3 Digit x 1 Multiplication & Division										
4 Digit x 1 Multiplication & Division										
2 Digit x 2 Multiplication & Division										
3 Digit x 2 Multiplication & Division										
4 Digit x 2 Multiplication & Division										

GOALS										
3 Digit x 3 Multiplication & Division										
4 Digit x 3 Multiplication & Division										
2 Digit x 3 Multiplication & Division										
4 Digit x 4 Multiplication & Division										

FOR INSTRUCTOR USE ONLY

Table of Contents

Introduction .... p. i
Group Grids & Student Mastery Forms .... p. vii
Unit 1
Addition & Subtraction .... p. 1 - 2
1. 1 Digit + 1 Digit Addition & Subtraction .... p. 3
Place Value .... p. 10
Regrouping .... p. 11
Borrowing .... p. 12
2. 2 Digit + 1 Digit Addition & Subtraction .... p. 13
3. 3 Digit + 1 Digit Addition & Subtraction .... p. 20
4. 4 Digit + 1 Digit Addition & Subtraction .... p. 26
5. 2 Digit + 2 Digit Addition & Subtraction .... p. 33
6. 3 Digit + 2 Digit Addition & Subtraction .... p. 40
7. 4 Digit + 2 Digit Addition & Subtraction .... p. 47
8. 3 Digit + 3 Digit Addition & Subtraction .... p. 54
9. 4 Digit + 3 Digit Addition & Subtraction .... p. 61
10. 4 Digit + 4 Digit Addition & Subtraction .... p. 68
Unit 2
Multiplication & Division .... p. 75 - 76
11. 1 Digit x 1 Multiplication & Division .... p. 77
12. 2 Digit x 1 Multiplication & Division .... p. 84
13. 3 Digit x 1 Multiplication & Division .... p. 91
14. 4 Digit x 1 Multiplication & Division .... p. 98
15. 2 Digit x 2 Multiplication & Division .... p. 105
16. 3 Digit x 2 Multiplication & Division .... p. 112
17. 4 Digit x 2 Multiplication & Division .... p. 119
18. 3 Digit x 3 Multiplication & Division .... p. 126
19. 4 Digit x 3 Multiplication & Division .... p. 133
20. 4 Digit x 4 Multiplication & Division .... p. 140
What's Next? .... p. I

Addition & Subtraction

This section introduces you to ways to count numbers and objects. The first method is called addition.

Addition is represented by this symbol: +

ADDITION

Addition is the practice of counting on. You need to count on to see how many of something you have in all. For instance, Dr. Jones wants to know how many girls and how many boys turned in their homework assignments.

1 + 2 = 3

To solve this problem, we could organize the students into girls and boys and then count them. Or we could let numbers represent them and add them all together.

One girl plus two boys equal three students in all.

Addition & Subtraction

We have already learned about addition. This section also introduces you to ways to count numbers and objects. The next method is called subtraction.

Subtraction is represented by this symbol: -

SUBTRACTION

Subtraction is the practice of taking away. You need to take away to see how many of something is gone. For instance, Dr. Jones wants to know how students are left if one goes home early.

3 - 1 = 2

To solve this problem, we will start with the number of students in all then take away the one who goes home. We will count how many are left and that will be our answer. Three take away one is two.

1 DIGIT + 1 DIGIT ADDITION & SUBTRACTION

SITUATION: Dr. Jones has two delicious popsicles. She brought one more delicious popsicle. How many popsicles does she have in all?

Answer: Two popsicles plus one popsicle equals three popsicles

Show with numbers 2 + 1 = 3

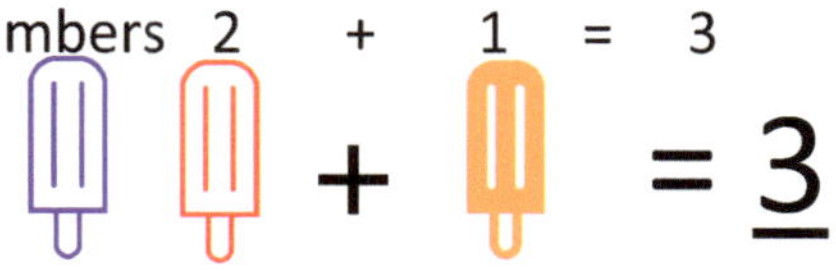

1. 1 + 1	**2.** 2 + 2	**3.** 3 + 3	**4.** 4 + 4	**5.** 5 + 1
6. 7 + 0	**7.** 3 + 4	**8.** 5 + 3	**9.** 1 + 0	**10.** 5 + 4
11. 5 + 2	**12.** 4 + 5	**13.** 3 + 2	**14.** 5 + 0	**15.** 7 + 2

LESSON 1
1 DIGIT + 1 DIGIT ADDITION & SUBTRACTION

$$\begin{array}{r} 8 \\ -\ 6 \\ \hline 2 \end{array}$$

1. $\begin{array}{r} 5 \\ -\ 3 \\ \hline \end{array}$	**2.** $\begin{array}{r} 4 \\ -\ 3 \\ \hline \end{array}$	**3.** $\begin{array}{r} 2 \\ -\ 0 \\ \hline \end{array}$	**4.** $\begin{array}{r} 5 \\ -\ 3 \\ \hline \end{array}$	**5.** $\begin{array}{r} 6 \\ -\ 3 \\ \hline \end{array}$
6. $\begin{array}{r} 7 \\ -\ 6 \\ \hline \end{array}$	**7.** $\begin{array}{r} 8 \\ -\ 1 \\ \hline \end{array}$	**8.** $\begin{array}{r} 4 \\ -\ 3 \\ \hline \end{array}$	**9.** $\begin{array}{r} 6 \\ -\ 5 \\ \hline \end{array}$	**10.** $\begin{array}{r} 7 \\ -\ 3 \\ \hline \end{array}$
11. $\begin{array}{r} 8 \\ -\ 4 \\ \hline \end{array}$	**12.** $\begin{array}{r} 9 \\ -\ 5 \\ \hline \end{array}$	**13.** $\begin{array}{r} 6 \\ -\ 4 \\ \hline \end{array}$	**14.** $\begin{array}{r} 5 \\ -\ 3 \\ \hline \end{array}$	**15.** $\begin{array}{r} 7 \\ -\ 5 \\ \hline \end{array}$

LESSON 1- REVIEW A+
1 DIGIT + 1 DIGIT ADDITION & SUBTRACTION

$$\begin{array}{r} 8 \\ +\ 6 \\ \hline 14 \end{array}$$

1. 4 + 2	2. 3 + 5	3. 7 + 1	4. 4 + 3	5. 6 + 2
6. 6 + 2	7. 7 + 1	8. 6 + 3	9. 4 + 4	10. 5 + 2
11. 5 + 1	12. 9 + 0	13. 5 + 3	14. 8 + 1	15. 0 + 0

LESSON 1 – REVIEW A -
1 DIGIT + 1 DIGIT ADDITION & SUBTRACTION

$$\begin{array}{r} 9 \\ \underline{-\ 4} \end{array}$$

1. $\begin{array}{r} 9 \\ \underline{-\ 6} \end{array}$	**2.** $\begin{array}{r} 7 \\ \underline{-\ 5} \end{array}$	**3.** $\begin{array}{r} 7 \\ \underline{-\ 2} \end{array}$	**4.** $\begin{array}{r} 9 \\ \underline{-\ 5} \end{array}$	**5.** $\begin{array}{r} 6 \\ \underline{-\ 2} \end{array}$
6. $\begin{array}{r} 8 \\ \underline{-\ 3} \end{array}$	**7.** $\begin{array}{r} 9 \\ \underline{-\ 6} \end{array}$	**8.** $\begin{array}{r} 4 \\ \underline{-\ 3} \end{array}$	**9.** $\begin{array}{r} 6 \\ \underline{-\ 4} \end{array}$	**10.** $\begin{array}{r} 5 \\ \underline{-\ 2} \end{array}$
11. $\begin{array}{r} 7 \\ \underline{-\ 6} \end{array}$	**12.** $\begin{array}{r} 3 \\ \underline{-\ 2} \end{array}$	**13.** $\begin{array}{r} 2 \\ \underline{-\ 2} \end{array}$	**14.** $\begin{array}{r} 4 \\ \underline{-\ 1} \end{array}$	**15.** $\begin{array}{r} 6 \\ \underline{-\ 6} \end{array}$

LESSON 1 REVIEW B +
1 DIGIT + 1 DIGIT ADDITION & SUBTRACTION

$$\begin{array}{r} 7 \\ +\ 6 \\ \hline \end{array}$$

1. $\begin{array}{r} 6 \\ +\ 6 \\ \hline \end{array}$	2. $\begin{array}{r} 3 \\ +\ 7 \\ \hline \end{array}$	3. $\begin{array}{r} 4 \\ +\ 7 \\ \hline \end{array}$	4. $\begin{array}{r} 5 \\ +\ 3 \\ \hline \end{array}$	5. $\begin{array}{r} 4 \\ +\ 1 \\ \hline \end{array}$
6. $\begin{array}{r} 3 \\ +\ 2 \\ \hline \end{array}$	7. $\begin{array}{r} 4 \\ +\ 3 \\ \hline \end{array}$	8. $\begin{array}{r} 2 \\ +\ 7 \\ \hline \end{array}$	9. $\begin{array}{r} 3 \\ +\ 9 \\ \hline \end{array}$	10. $\begin{array}{r} 6 \\ +\ 3 \\ \hline \end{array}$
11. $\begin{array}{r} 4 \\ +\ 8 \\ \hline \end{array}$	12. $\begin{array}{r} 9 \\ +\ 4 \\ \hline \end{array}$	13. $\begin{array}{r} 1 \\ +\ 7 \\ \hline \end{array}$	14. $\begin{array}{r} 6 \\ +\ 5 \\ \hline \end{array}$	15. $\begin{array}{r} 7 \\ +\ 9 \\ \hline \end{array}$

LESSON 1 REVIEW B –
1 DIGIT - 1 DIGIT ADDITION & SUBTRACTION

$$\begin{array}{r} 7 \\ -\ 6 \\ \hline \end{array}$$

1. $\begin{array}{r} 9 \\ -\ \ 9 \\ \hline \end{array}$	2. $\begin{array}{r} 8 \\ -\ \ 8 \\ \hline \end{array}$	3. $\begin{array}{r} 7 \\ -\ \ 7 \\ \hline \end{array}$	4. $\begin{array}{r} 6 \\ -\ \ 6 \\ \hline \end{array}$	5. $\begin{array}{r} 5 \\ -\ \ 5 \\ \hline \end{array}$
6. $\begin{array}{r} 9 \\ -\ 6 \\ \hline \end{array}$	7. $\begin{array}{r} 9 \\ -\ 5 \\ \hline \end{array}$	8. $\begin{array}{r} 9 \\ -\ 7 \\ \hline \end{array}$	9. $\begin{array}{r} 9 \\ -\ \ 8 \\ \hline \end{array}$	10. $\begin{array}{r} 9 \\ -\ \ 9 \\ \hline \end{array}$
11. $\begin{array}{r} 8 \\ -\ \ 7 \\ \hline \end{array}$	12. $\begin{array}{r} 8 \\ -\ \ 6 \\ \hline \end{array}$	13. $\begin{array}{r} 8 \\ -\ \ 5 \\ \hline \end{array}$	14. $\begin{array}{r} 8 \\ -\ \ 4 \\ \hline \end{array}$	15. $\begin{array}{r} 8 \\ -\ 3 \\ \hline \end{array}$

1 DIGIT SPEED TEST

1. 9 + 7	2. 3 + 4	3. 1 + 5	4. 2 + 2	5. 5 + 7
6. 5 - 3	7. 9 - 8	8. 7 - 1	9. 4 - 2	10. 9 - 1
11. 8 + 7	12. 8 + 2	13. 2 - 4	14. 6 - 9	15. 8 - 2
16. 2 - 1	17. 8 - 6	18. 4 - 3	19. 3 + 2	20. 4 - 4

QUICK LESSON 1: PLACE VALUE

CONNECT: In Lesson 1, we learned to add a one-digit number to another one-digit number, like 4 + 3 + 7. In the following Lessons, we will add larger numbers.

LEARN: When adding and subtracting a two-digit number to a one-digit number, you need to understand WHERE to write your answer.

QUICK LEARN: **Place Value** – the value a digit has because of its place in a number.

PLACE VALUE CHART

	tens	ones
	1	3
add +		2
total	**1**	**5**

When adding a two-digit number like 13 to a one-digit number like 2, FIRST be sure the problem is lined up correctly with the <u>ones on top of the ones</u> and <u>tens on top of the tens.</u>

SEE: 13
+ 2

• Cover the tens column while you add the ones together. Next, uncover the tens column and add the tens. What is your answer?

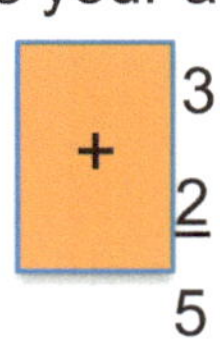

Three plus two is five.

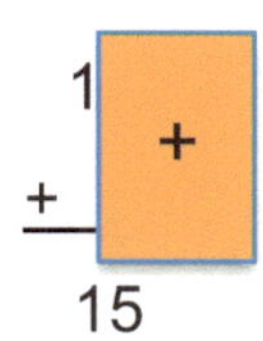

One plus nothing is one.

13 + 2 = 15
Thirteen plus two is fifteen.

QUICK LESSON 2: REGROUPING with Addition

CONNECT: In Quick Lesson 1, we learned that only one digit can fit into each place value box. In the following Lessons, we will add larger numbers that will change how we manage our place value.

LEARN: When adding and subtracting multi-digit numbers, sometimes your answer will have more than one digit. Only one digit can fit into each place value box, so we will place the one digit in the box and put the tens digit at the top of the next column. This practice is called regrouping.

QUICK LEARN: **Regrouping** – the extra digit must be placed on top because it will not fit into the place value box.

PLACE VALUE CHART

	tens	ones
	1	5
add +		5
total	**2**	**0**

Place the 0 in the ones place. Write the 1 on top of the tens place box.

When adding a two-digit number like 15 to a one-digit number like 5, FIRST be sure the problem is lined up correctly with the ones on top of the ones and tens on top of the tens.

SEE: 15

+ 5

• Cover the tens column while you add the ones together. In this problem, your answer to the one's column has two digits. So, we will write the number that fits in the tens place at the top and add it in when we move to the tens column to do our addition. Next, uncover the tens column and add the tens. In the tens place, we have 1+1 = 2. What is your answer? 15 + 5 = 20

QUICK LESSON 3: BORROWING with Subtraction

CONNECT: In Quick Lesson 2, we learned that only one digit can fit into each place value box, so we regrouped by moving the digit that belongs to the tens place to the top of its column. In this lesson, we will learn to manage place value in subtraction.

LEARN: When subtracting multi-digit numbers sometimes your **minuend** – the number that you subtract another number from - will be smaller than your **subtrahend** – the number that you subtract from the minuend. The answer to a subtraction problem is called the **Difference.** When this happens, you will have to do a special task called borrowing.

QUICK LEARN: **Borrowing –** when the top number is too small to subtract from, you "go next door and borrow one more" from the number next to it. Then you change the values and subtract.

PLACE VALUE CHART

	tens	ones
	~~2~~ 1	15 ~~5~~
subtract -		7
total	1	8

Borrow 1 from the 2. Change the 5 to 15 and the 2 to 1. Subtract.

When subtracting a number like 7 from 25, NOTICE that the 5 is too small to take 7 out of it, so I will go next door and borrow one more from the 2. When I borrow a 1, the 5 becomes 15 and the 2 becomes a 1.

SEE: 25

$\underline{-\ \ 7}$

- Next, subtract 7 from 15 to get 8. Then bring down the 1 or subtract 0 from 1 to get 1. Your final answer is 18.

LESSON 2 +
2 DIGIT + 1 DIGIT ADDITION & SUBTRACTION

4
+ 2

1. 1 + 1	**2.** 2 +2	**3.** 3 + 3	**4.** 4 + 4	**5.** 5 + 1
6. 7 + 0	**7.** 3 + 4	**8.** 5 + 3	**9.** 1 + 0	**10.** 5 + 4
11. 5 + 2	**12.** 4 + 5	**13.** 3 + 2	**14.** 5 +0	**15.** 7 + 2

LESSON 2 –
2 DIGIT + 1 DIGIT ADDITION & SUBTRACTION

$$\begin{array}{r} 24 \\ -\ 6 \\ \hline 18 \end{array}$$

1. $\begin{array}{r} 12 \\ -\ 3 \\ \hline \end{array}$	2. $\begin{array}{r} 10 \\ -\ 3 \\ \hline \end{array}$	3. $\begin{array}{r} 15 \\ -\ 9 \\ \hline \end{array}$	4. $\begin{array}{r} 22 \\ -\ 6 \\ \hline \end{array}$	5. $\begin{array}{r} 11 \\ -\ 9 \\ \hline \end{array}$
6. $\begin{array}{r} 17 \\ -\ 3 \\ \hline \end{array}$	7. $\begin{array}{r} 18 \\ -\ 9 \\ \hline \end{array}$	8. $\begin{array}{r} 22 \\ -\ 2 \\ \hline \end{array}$	9. $\begin{array}{r} 17 \\ -\ 7 \\ \hline \end{array}$	10. $\begin{array}{r} 20 \\ -\ 3 \\ \hline \end{array}$
11. $\begin{array}{r} 14 \\ -\ 4 \\ \hline \end{array}$	12. $\begin{array}{r} 15 \\ -\ 9 \\ \hline \end{array}$	13. $\begin{array}{r} 23 \\ -\ 6 \\ \hline \end{array}$	14. $\begin{array}{r} 50 \\ -\ 5 \\ \hline \end{array}$	15. $\begin{array}{r} 66 \\ -\ 3 \\ \hline \end{array}$

LESSON 2 REVIEW A +
2 DIGIT + 1 DIGIT ADDITION & SUBTRACTION

$$\begin{array}{r} 59 \\ \underline{+\ 4} \end{array}$$

1. $\begin{array}{r} 24 \\ \underline{+\ \ 5} \end{array}$	2. $\begin{array}{r} 35 \\ \underline{+\ \ 3} \end{array}$	3. $\begin{array}{r} 44 \\ \underline{+\ 4} \end{array}$	4. $\begin{array}{r} 52 \\ \underline{+\ 3} \end{array}$	5. $\begin{array}{r} 69 \\ \underline{+\ 1} \end{array}$
6. $\begin{array}{r} 56 \\ \underline{+\ 2} \end{array}$	7. $\begin{array}{r} 20 \\ \underline{+\ 8} \end{array}$	8. $\begin{array}{r} 36 \\ \underline{+\ 2} \end{array}$	9. $\begin{array}{r} 62 \\ \underline{+\ 5} \end{array}$	10. $\begin{array}{r} 17 \\ \underline{+\ 2} \end{array}$
11. $\begin{array}{r} 41 \\ \underline{+\ 8} \end{array}$	12. $\begin{array}{r} 52 \\ \underline{+\ 5} \end{array}$	13. $\begin{array}{r} 62 \\ \underline{+\ 4} \end{array}$	14. $\begin{array}{r} 73 \\ \underline{+\ 5} \end{array}$	15. $\begin{array}{r} 25 \\ \underline{+\ 4} \end{array}$

LESSON 2 REVIEW A -
2 DIGIT + 1 DIGIT ADDITION & SUBTRACTION

$$\begin{array}{r} 43 \\ -\ 5 \\ \hline \end{array}$$

1. $\begin{array}{r} 24 \\ -\ 5 \\ \hline \end{array}$	2. $\begin{array}{r} 35 \\ -\ 3 \\ \hline \end{array}$	3. $\begin{array}{r} 44 \\ -\ 4 \\ \hline \end{array}$	4. $\begin{array}{r} 52 \\ -\ 3 \\ \hline \end{array}$	5. $\begin{array}{r} 69 \\ -\ 1 \\ \hline \end{array}$
6. $\begin{array}{r} 56 \\ -\ 2 \\ \hline \end{array}$	7. $\begin{array}{r} 20 \\ -\ 8 \\ \hline \end{array}$	8. $\begin{array}{r} 36 \\ -\ 2 \\ \hline \end{array}$	9. $\begin{array}{r} 62 \\ -\ 5 \\ \hline \end{array}$	10. $\begin{array}{r} 17 \\ -\ 2 \\ \hline \end{array}$
11. $\begin{array}{r} 41 \\ -\ 8 \\ \hline \end{array}$	12. $\begin{array}{r} 52 \\ -\ 5 \\ \hline \end{array}$	13. $\begin{array}{r} 62 \\ -\ 4 \\ \hline \end{array}$	14. $\begin{array}{r} 73 \\ -\ 5 \\ \hline \end{array}$	15. $\begin{array}{r} 25 \\ -\ 4 \\ \hline \end{array}$

LESSON 2 REVIEW B +
2 DIGIT + 1 DIGIT ADDITION & SUBTRACTION

$$\begin{array}{r} 15 \\ +\ 6 \\ \hline \end{array}$$

1. $\begin{array}{r} 93 \\ +\ 4 \\ \hline \end{array}$	2. $\begin{array}{r} 80 \\ +\ 3 \\ \hline \end{array}$	3. $\begin{array}{r} 6 \\ +\ 3 \\ \hline \end{array}$	4. $\begin{array}{r} 73 \\ +\ 4 \\ \hline \end{array}$	5. $\begin{array}{r} 53 \\ +\ 5 \\ \hline \end{array}$
6. $\begin{array}{r} 62 \\ +\ 5 \\ \hline \end{array}$	7. $\begin{array}{r} 53 \\ +\ 4 \\ \hline \end{array}$	8. $\begin{array}{r} 7 \\ +\ 2 \\ \hline \end{array}$	9. $\begin{array}{r} 11 \\ +\ 1 \\ \hline \end{array}$	10. $\begin{array}{r} 16 \\ +\ 1 \\ \hline \end{array}$
11. $\begin{array}{r} 46 \\ +\ 3 \\ \hline \end{array}$	12. $\begin{array}{r} 55 \\ +\ 3 \\ \hline \end{array}$	13. $\begin{array}{r} 64 \\ +\ 5 \\ \hline \end{array}$	14. $\begin{array}{r} 81 \\ +\ 6 \\ \hline \end{array}$	15. $\begin{array}{r} 26 \\ +\ 2 \\ \hline \end{array}$

LESSON 2 REVIEW B –
2 DIGIT + 1 DIGIT ADDITION & SUBTRACTION

$$\begin{array}{r} 10 \\ -\ 6 \\ \hline \end{array}$$

1. $\begin{array}{r} 4 \\ -\ 3 \\ \hline \end{array}$	2. $\begin{array}{r} 6 \\ -\ 3 \\ \hline \end{array}$	3. $\begin{array}{r} 10 \\ -\ 7 \\ \hline \end{array}$	4. $\begin{array}{r} 15 \\ -\ 6 \\ \hline \end{array}$	5. $\begin{array}{r} 20 \\ -\ 5 \\ \hline \end{array}$
6. $\begin{array}{r} 56 \\ -\ 3 \\ \hline \end{array}$	7. $\begin{array}{r} 48 \\ -\ 8 \\ \hline \end{array}$	8. $\begin{array}{r} 54 \\ -\ 9 \\ \hline \end{array}$	9. $\begin{array}{r} 14 \\ -\ 6 \\ \hline \end{array}$	10. $\begin{array}{r} 58 \\ -\ 6 \\ \hline \end{array}$
11. $\begin{array}{r} 18 \\ -\ 6 \\ \hline \end{array}$	12. $\begin{array}{r} 13 \\ -\ 5 \\ \hline \end{array}$	13. $\begin{array}{r} 40 \\ -\ 9 \\ \hline \end{array}$	14. $\begin{array}{r} 55 \\ -\ 6 \\ \hline \end{array}$	15. $\begin{array}{r} 30 \\ -\ 6 \\ \hline \end{array}$

2/1 DIGIT SPEED TEST

1. 87 + 6	2. 48 - 6	3. 19 + 4	4. 17 + 9	5. 36 - 7
6. 58 - 9	7. 91 - 7	8. 54 - 8	9. 43 + 9	10. 91 - 6
11. 47 + 7	12. 36 + 2	13. 17 - 4	14. 56 - 9	15. 47 - 8
16. 12 - 6	17. 34 - 9	18. 14 - 6	19. 98 + 8	20. 48 - 9

LESSON 3 +
3 DIGIT + 1 DIGIT ADDITION & SUBTRACTION

$$\begin{array}{r} 123 \\ +\ \ 1 \\ \hline 124 \end{array}$$

1. $\begin{array}{r} 111 \\ +\ \ 1 \\ \hline \end{array}$	2. $\begin{array}{r} 112 \\ +\ \ 3 \\ \hline \end{array}$	3. $\begin{array}{r} 123 \\ +\ \ 3 \\ \hline \end{array}$	4. $\begin{array}{r} 133 \\ +\ \ 3 \\ \hline \end{array}$	5. $\begin{array}{r} 140 \\ +\ \ 4 \\ \hline \end{array}$
6. $\begin{array}{r} 171 \\ +\ \ 3 \\ \hline \end{array}$	7. $\begin{array}{r} 190 \\ +\ \ 8 \\ \hline \end{array}$	8. $\begin{array}{r} 144 \\ +\ \ 4 \\ \hline \end{array}$	9. $\begin{array}{r} 199 \\ +\ \ 1 \\ \hline \end{array}$	10. $\begin{array}{r} 176 \\ +\ \ 3 \\ \hline \end{array}$
11. $\begin{array}{r} 111 \\ +\ \ 7 \\ \hline \end{array}$	12. $\begin{array}{r} 125 \\ +\ \ 5 \\ \hline \end{array}$	13. $\begin{array}{r} 100 \\ +\ \ 6 \\ \hline \end{array}$	14. $\begin{array}{r} 154 \\ +\ \ 3 \\ \hline \end{array}$	15. $\begin{array}{r} 160 \\ +\ \ 7 \\ \hline \end{array}$

LESSON 3 -
3 DIGIT + 1 DIGIT ADDITION & SUBTRACTION

$$\begin{array}{r} 123 \\ -\ 6 \\ \hline 117 \end{array}$$

1. $\begin{array}{r} 112 \\ -\ \ 3 \\ \hline \end{array}$	2. $\begin{array}{r} 114 \\ -\ \ 7 \\ \hline \end{array}$	3. $\begin{array}{r} 212 \\ -\ \ 6 \\ \hline \end{array}$	4. $\begin{array}{r} 411 \\ -\ \ 3 \\ \hline \end{array}$	5. $\begin{array}{r} 617 \\ -\ \ 9 \\ \hline \end{array}$
6. $\begin{array}{r} 407 \\ -\ \ 8 \\ \hline \end{array}$	7. $\begin{array}{r} 912 \\ -\ \ 7 \\ \hline \end{array}$	8. $\begin{array}{r} 414 \\ -\ \ 6 \\ \hline \end{array}$	9. $\begin{array}{r} 693 \\ -\ \ 8 \\ \hline \end{array}$	10. $\begin{array}{r} 456 \\ -\ \ 7 \\ \hline \end{array}$
11. $\begin{array}{r} 117 \\ -\ \ 8 \\ \hline \end{array}$	12. $\begin{array}{r} 290 \\ -\ \ 7 \\ \hline \end{array}$	13. $\begin{array}{r} 459 \\ -\ \ 6 \\ \hline \end{array}$	14. $\begin{array}{r} 593 \\ -\ \ 7 \\ \hline \end{array}$	15. $\begin{array}{r} 611 \\ -\ \ 9 \\ \hline \end{array}$

LESSON 3 REVIEW A –
3 DIGIT + 1 DIGIT ADDITION & SUBTRACTION

440
- 5

1. 633 − 7	2. 766 − 9	3. 313 − 8	4. 406 − 9	5. 503 − 7
6. 702 − 8	7. 601 − 5	8. 602 − 3	9. 907 − 8	10. 605 − 8
11. 611 − 8	12. 706 − 8	13. 411 − 3	14. 508 − 9	15. 612 − 3

LESSON 3 REVIEW A +
3 DIGIT + 1 DIGIT ADDITION & SUBTRACTION

$$\begin{array}{r} 704 \\ +\ \ 8 \\ \hline \end{array}$$

1. $\begin{array}{r} 444 \\ +\ \ 4 \\ \hline \end{array}$	2. $\begin{array}{r} 276 \\ +\ \ 3 \\ \hline \end{array}$	3. $\begin{array}{r} 603 \\ +\ \ 5 \\ \hline \end{array}$	4. $\begin{array}{r} 511 \\ +\ \ 1 \\ \hline \end{array}$	5. $\begin{array}{r} 603 \\ +\ \ 5 \\ \hline \end{array}$
6. $\begin{array}{r} 713 \\ +\ \ 5 \\ \hline \end{array}$	7. $\begin{array}{r} 804 \\ +\ \ 3 \\ \hline \end{array}$	8. $\begin{array}{r} 533 \\ +\ \ 6 \\ \hline \end{array}$	9. $\begin{array}{r} 217 \\ +\ \ 2 \\ \hline \end{array}$	10. $\begin{array}{r} 463 \\ +\ \ 6 \\ \hline \end{array}$
11. $\begin{array}{r} 407 \\ +\ \ 2 \\ \hline \end{array}$	12. $\begin{array}{r} 680 \\ +\ \ 7 \\ \hline \end{array}$	13. $\begin{array}{r} 413 \\ +\ \ 3 \\ \hline \end{array}$	14. $\begin{array}{r} 513 \\ +\ \ 2 \\ \hline \end{array}$	15. $\begin{array}{r} 603 \\ +\ \ 4 \\ \hline \end{array}$

LESSON 3 REVIEW B +
3 DIGIT + 1 DIGIT ADDITION & SUBTRACTION

179
+ 8

1. 205 + 3	2. 176 + 3	3. 156 + 3	4. 307 + 2	5. 611 + 5
6. 298 + 1	7. 342 + 7	8. 512 + 7	9. 673 + 4	10. 703 + 5
11. 711 + 8	12. 404 + 5	13. 999 + 1	14. 500 + 6	15. 756 + 3

3/1 DIGIT SPEED TEST

1. 337 + 7	2. 443 + 4	3. 971 + 5	4. 617 + 2	5. 498 + 7
6. 592 - 3	7. 993 - 8	8. 437 - 8	9. 847 - 2	10. 368 - 9
11. 813 + 7	12. 453 + 2	13. 785 - 6	14. 605 - 9	15. 876 - 8
16. 214 - 5	17. 118 - 6	18. 407 - 9	19. 317 + 2	20. 409 - 4

LESSON 4 +
4 DIGIT + 1 DIGIT ADDITION & SUBTRACTION

$$\begin{array}{r} 2430 \\ +\quad 7 \\ \hline 2437 \end{array}$$

1. $\begin{array}{r} 1336 \\ +\quad 2 \\ \hline \end{array}$	2. $\begin{array}{r} 1045 \\ +\quad 4 \\ \hline \end{array}$	3. $\begin{array}{r} 6760 \\ +\quad 6 \\ \hline \end{array}$	4. $\begin{array}{r} 1154 \\ +\quad 1 \\ \hline \end{array}$	5. $\begin{array}{r} 1222 \\ +\quad 9 \\ \hline \end{array}$
6. $\begin{array}{r} 4314 \\ +\quad 5 \\ \hline \end{array}$	7. $\begin{array}{r} 9212 \\ +\quad 6 \\ \hline \end{array}$	8. $\begin{array}{r} 5546 \\ +\quad 6 \\ \hline \end{array}$	9. $\begin{array}{r} 3150 \\ +\quad 9 \\ \hline \end{array}$	10. $\begin{array}{r} 2048 \\ +\quad 7 \\ \hline \end{array}$
11. $\begin{array}{r} 4976 \\ +\quad 5 \\ \hline \end{array}$	12. $\begin{array}{r} 3344 \\ +\quad 4 \\ \hline \end{array}$	13. $\begin{array}{r} 1448 \\ +\quad 3 \\ \hline \end{array}$	14. $\begin{array}{r} 1674 \\ +\quad 1 \\ \hline \end{array}$	15. $\begin{array}{r} 2293 \\ +\quad 8 \\ \hline \end{array}$

LESSON 4 –
4 DIGIT + 1 DIGIT ADDITION & SUBTRACTION

$$\begin{array}{r} 5880 \\ - \quad 5 \\ \hline 5875 \end{array}$$

1. $\begin{array}{r} 1234 \\ - \quad 9 \\ \hline \end{array}$	2. $\begin{array}{r} 4653 \\ - \quad 8 \\ \hline \end{array}$	3. $\begin{array}{r} 3932 \\ - \quad 6 \\ \hline \end{array}$	4. $\begin{array}{r} 5893 \\ - \quad 7 \\ \hline \end{array}$	5. $\begin{array}{r} 7633 \\ - \quad 2 \\ \hline \end{array}$
6. $\begin{array}{r} 7633 \\ - \quad 7 \\ \hline \end{array}$	7. $\begin{array}{r} 4849 \\ - \quad 3 \\ \hline \end{array}$	8. $\begin{array}{r} 7878 \\ - \quad 9 \\ \hline \end{array}$	9. $\begin{array}{r} 1407 \\ - \quad 9 \\ \hline \end{array}$	10. $\begin{array}{r} 1224 \\ - \quad 7 \\ \hline \end{array}$
11. $\begin{array}{r} 4303 \\ - \quad 7 \\ \hline \end{array}$	12. $\begin{array}{r} 4809 \\ - \quad 7 \\ \hline \end{array}$	13. $\begin{array}{r} 1444 \\ - \quad 6 \\ \hline \end{array}$	14. $\begin{array}{r} 1508 \\ - \quad 8 \\ \hline \end{array}$	15. $\begin{array}{r} 1507 \\ - \quad 8 \\ \hline \end{array}$

LESSON 4 REVIEW A +
4 DIGIT + 1 DIGIT ADDITION & SUBTRACTION

$$\begin{array}{r} 5500 \\ +\quad 8 \\ \hline 5508 \end{array}$$

1. $\begin{array}{r} 5324 \\ +\quad 5 \\ \hline \end{array}$	2. $\begin{array}{r} 4635 \\ +\quad 3 \\ \hline \end{array}$	3. $\begin{array}{r} 5144 \\ +\quad 4 \\ \hline \end{array}$	4. $\begin{array}{r} 5276 \\ +\quad 3 \\ \hline \end{array}$	5. $\begin{array}{r} 6519 \\ +\quad 1 \\ \hline \end{array}$
6. $\begin{array}{r} 298 \\ +\ 1 \\ \hline \end{array}$	7. $\begin{array}{r} 342 \\ +\ 7 \\ \hline \end{array}$	8. $\begin{array}{r} 512 \\ +\ 7 \\ \hline \end{array}$	9. $\begin{array}{r} 673 \\ +\ 4 \\ \hline \end{array}$	10. $\begin{array}{r} 703 \\ +\ 5 \\ \hline \end{array}$
11. $\begin{array}{r} 711 \\ +\ 8 \\ \hline \end{array}$	12. $\begin{array}{r} 404 \\ +\ 5 \\ \hline \end{array}$	13. $\begin{array}{r} 999 \\ +\ 1 \\ \hline \end{array}$	14. $\begin{array}{r} 500 \\ +\ 6 \\ \hline \end{array}$	15. $\begin{array}{r} 756 \\ +\ 3 \\ \hline \end{array}$

LESSON 4 REVIEW A -
4 DIGIT + 1 DIGIT ADDITION & SUBTRACTION

$$\begin{array}{r} 1000 \\ - \quad 7 \\ \hline 993 \end{array}$$

1. $\begin{array}{r} 1706 \\ - \quad 8 \\ \hline \end{array}$	2. $\begin{array}{r} 1705 \\ - \quad 7 \\ \hline \end{array}$	3. $\begin{array}{r} 1212 \\ - \quad 8 \\ \hline \end{array}$	4. $\begin{array}{r} 4038 \\ - \quad 9 \\ \hline \end{array}$	5. $\begin{array}{r} 1111 \\ - \quad 8 \\ \hline \end{array}$
6. $\begin{array}{r} 1207 \\ - \quad 9 \\ \hline \end{array}$	7. $\begin{array}{r} 4058 \\ - \quad 9 \\ \hline \end{array}$	8. $\begin{array}{r} 1403 \\ - \quad 8 \\ \hline \end{array}$	9. $\begin{array}{r} 1212 \\ - \quad 7 \\ \hline \end{array}$	10. $\begin{array}{r} 1405 \\ - \quad 7 \\ \hline \end{array}$
11. $\begin{array}{r} 1111 \\ - \quad 5 \\ \hline \end{array}$	12. $\begin{array}{r} 1403 \\ - \quad 9 \\ \hline \end{array}$	13. $\begin{array}{r} 7706 \\ - \quad 7 \\ \hline \end{array}$	14. $\begin{array}{r} 5503 \\ - \quad 6 \\ \hline \end{array}$	15. $\begin{array}{r} 1901 \\ - \quad 3 \\ \hline \end{array}$

LESSON 4 REVIEW B +
4 DIGIT + 1 DIGIT ADDITION & SUBTRACTION

$$\begin{array}{r} 2234 \\ +\quad 8 \\ \hline 2242 \end{array}$$

1. $\begin{array}{r} 9321 \\ +\quad 7 \\ \hline \end{array}$	2. $\begin{array}{r} 8028 \\ +\quad 6 \\ \hline \end{array}$	3. $\begin{array}{r} 450 \\ +\quad 7 \\ \hline \end{array}$	4. $\begin{array}{r} 7305 \\ +\quad 8 \\ \hline \end{array}$	5. $\begin{array}{r} 5302 \\ +\quad 5 \\ \hline \end{array}$
6. $\begin{array}{r} 6211 \\ +\quad 5 \\ \hline \end{array}$	7. $\begin{array}{r} 5312 \\ +\quad 4 \\ \hline \end{array}$	8. $\begin{array}{r} 7151 \\ +\quad 2 \\ \hline \end{array}$	9. $\begin{array}{r} 1120 \\ +\quad 1 \\ \hline \end{array}$	10. $\begin{array}{r} 1650 \\ +\quad 1 \\ \hline \end{array}$
11. $\begin{array}{r} 4651 \\ +\quad 3 \\ \hline \end{array}$	12. $\begin{array}{r} 5560 \\ +\quad 3 \\ \hline \end{array}$	13. $\begin{array}{r} 6432 \\ +\quad 5 \\ \hline \end{array}$	14. $\begin{array}{r} 8125 \\ +\quad 6 \\ \hline \end{array}$	15. $\begin{array}{r} 2623 \\ +\quad 2 \\ \hline \end{array}$

LESSON 4 REVIEW B -
4 DIGIT + 1 DIGIT ADDITION & SUBTRACTION

$$\begin{array}{r} 4570 \\ -\quad 3 \\ \hline 4567 \end{array}$$

1. 3450 − 7	2. 2209 − 3	3. 7821 − 9	4. 4257 − 4	5. 1021 − 7
6. 2500 − 8	7. 4255 − 6	8. 2511 − 6	9. 8695 − 7	10. 1335 − 8
11. 1587 − 5	12. 1273 − 5	13. 8755 − 6	14. 5583 − 4	15. 1852 − 6

4/1 DIGIT SPEED TEST

1. 1973 + 8	2. 7603 + 8	3. 5683 - 7	4. 4756 + 4	5. 7793 + 9
6. 5924 - 8	7. 9214 - 7	8. 1006 - 8	9. 4200 - 4	10. 1350 - 9
11. 5504 + 6	12. 4120 - 7	13. 1500 - 7	14. 4700 - 9	15. 3750 - 8
16. 2722 - 5	17. 1407 - 9	18. 7250 - 9	19. 1300 + 2	20. 1300 - 4

LESSON 5 +
2 DIGIT + 2 DIGIT ADDITION & SUBTRACTION

$$\begin{array}{r} 85 \\ +\ 10 \\ \hline 95 \end{array}$$

1. 19 +18	2. 72 + 13	3. 44 + 45	4. 16 +19	5. 33 +45
6. 16 +12	7. 33 +27	8. 38 +19	9. 45 +73	10. 17 +38
11. 45 +22	12. 67 +15	13. 55 +38	14. 83 +66	15. 92 +78

LESSON 5 -

2 DIGIT + 2 DIGIT ADDITION & SUBTRACTION

$$\begin{array}{r} 85 \\ -\ 10 \\ \hline 75 \end{array}$$

1. 44 - 36	2. 78 - 45	3. 88 - 49	4. 76 -38	5. 17 -15
6. 44 -38	7. 98 -79	8. 84 -38	9. 77 -29	10. 62 -46
11. 81 -67	12. 33 -16	13. 41 -29	14. 64 -46	15. 91 -86

LESSON 5 REVIEW A +
2 DIGIT + 2 DIGIT ADDITION & SUBTRACTION

$$\begin{array}{r} 76 \\ +\ 10 \\ \hline 86 \end{array}$$

1. $\begin{array}{r} 14 \\ +\ 37 \\ \hline \end{array}$	2. $\begin{array}{r} 33 \\ +\ 16 \\ \hline \end{array}$	3. $\begin{array}{r} 15 \\ +15 \\ \hline \end{array}$	4. $\begin{array}{r} 60 \\ +\ 40 \\ \hline \end{array}$	5. $\begin{array}{r} 16 \\ +\ 17 \\ \hline \end{array}$
6. $\begin{array}{r} 12 \\ +3 \\ \hline \end{array}$	7. $\begin{array}{r} 39 \\ +12 \\ \hline \end{array}$	8. $\begin{array}{r} 14 \\ +17 \\ \hline \end{array}$	9. $\begin{array}{r} 40 \\ +\ 50 \\ \hline \end{array}$	10. $\begin{array}{r} 11 \\ +76 \\ \hline \end{array}$
11. $\begin{array}{r} 17 \\ +\ 36 \\ \hline \end{array}$	12. $\begin{array}{r} 33 \\ +\ 27 \\ \hline \end{array}$	13. $\begin{array}{r} 48 \\ +\ 14 \\ \hline \end{array}$	14. $\begin{array}{r} 58 \\ +\ 76 \\ \hline \end{array}$	15. $\begin{array}{r} 38 \\ +\ 3 \\ \hline \end{array}$

LESSON 5 - REVIEW A –
2 DIGIT + 2 DIGIT ADDITION & SUBTRACTION

89
- 20

1. 44 - 36	2. 33 - 27	3. 14 - 7	4. 53 -37	5. 46 -19
6. 38 -29	7. 42 -37	8. 44 -37	9. 53 -19	10. 60 -37
11. 14 - 8	12. 9 - 4	13. 100 - 47	14. 16 - 15	15. 37 - 19

LESSON 5 REVIEW B –
2 DIGIT + 2 DIGIT ADDITION & SUBTRACTION

$$\begin{array}{r} 76 \\ -\ 23 \\ \hline \end{array}$$

1. 56 − 17	**2.** 43 − 18	**3.** 43 − 16	**4.** 58 − 29	**5.** 66 − 48
6. 88 − 79	**7.** 54 − 39	**8.** 62 − 48	**9.** 55 − 37	**10.** 92 − 39
11. 46 − 37	**12.** 55 − 48	**13.** 12 − 9	**14.** 531 − 26	**15.** 71 − 59

LESSON 5 REVIEW B +
2 DIGIT + 2 DIGIT ADDITION & SUBTRACTION

$$\begin{array}{r} 31 \\ +\ 12 \\ \hline \end{array}$$

1. $\begin{array}{r} 19 \\ +\ 67 \\ \hline \end{array}$	2. $\begin{array}{r} 38 \\ +\ 19 \\ \hline \end{array}$	3. $\begin{array}{r} 16 \\ +\ 6 \\ \hline \end{array}$	4. $\begin{array}{r} 4 \\ +\ 9 \\ \hline \end{array}$	5. $\begin{array}{r} 100 \\ +\ 3 \\ \hline \end{array}$
6. $\begin{array}{r} 1458 \\ +\ 6 \\ \hline \end{array}$	7. $\begin{array}{r} 5450 \\ +\ 9 \\ \hline \end{array}$	8. $\begin{array}{r} 5046 \\ +\ 3 \\ \hline \end{array}$	9. $\begin{array}{r} 17 \\ +\ 16 \\ \hline \end{array}$	10. $\begin{array}{r} 97 \\ +\ 13 \\ \hline \end{array}$
11. $\begin{array}{r} 123 \\ +\ 11 \\ \hline \end{array}$	12. $\begin{array}{r} 454 \\ +\ 6 \\ \hline \end{array}$	13. $\begin{array}{r} 14 \\ +\ 15 \\ \hline \end{array}$	14. $\begin{array}{r} 26 \\ +\ 18 \\ \hline \end{array}$	15. $\begin{array}{r} 22 \\ +\ 24 \\ \hline \end{array}$

2/2 DIGIT SPEED TEST

1. 40 + 28	2. 50 + 31	3. 14 +12	4. 38 + 41	5. 52 + 20
6. 46 - 17	7. 96 - 18	8. 45 - 97	9. 62 - 47	10. 54 - 29
11. 33 + 40	12. 56 + 72	13. 50 - 26	14. 30 - 19	15. 87 - 28
16. 26 - 17	17. 40 - 16	18. 20 - 19	19. 31 + 17	20. 19 - 13

LESSON 6 +
3 DIGIT + 2 DIGIT ADDITION & SUBTRACTION

$$\begin{array}{r} 123 \\ +\ 11 \\ \hline 134 \end{array}$$

1. $\begin{array}{r} 176 \\ +\ 13 \\ \hline \end{array}$	2. $\begin{array}{r} 197 \\ +\ 12 \\ \hline \end{array}$	3. $\begin{array}{r} 400 \\ +\ 10 \\ \hline \end{array}$	4. $\begin{array}{r} 453 \\ +\ 13 \\ \hline \end{array}$	5. $\begin{array}{r} 667 \\ +\ 37 \\ \hline \end{array}$
6. $\begin{array}{r} 133 \\ +\ 19 \\ \hline \end{array}$	7. $\begin{array}{r} 114 \\ +\ 15 \\ \hline \end{array}$	8. $\begin{array}{r} 190 \\ +\ 18 \\ \hline \end{array}$	9. $\begin{array}{r} 170 \\ +\ 30 \\ \hline \end{array}$	10. $\begin{array}{r} 153 \\ +\ 17 \\ \hline \end{array}$
11. $\begin{array}{r} 233 \\ +\ 48 \\ \hline \end{array}$	12. $\begin{array}{r} 443 \\ +\ 12 \\ \hline \end{array}$	13. $\begin{array}{r} 436 \\ +\ 36 \\ \hline \end{array}$	14. $\begin{array}{r} 676 \\ +\ 76 \\ \hline \end{array}$	15. $\begin{array}{r} 513 \\ +\ 19 \\ \hline \end{array}$

LESSON 6 –
3 DIGIT + 2 DIGIT ADDITION & SUBTRACTION

$$\begin{array}{r} 123 \\ -\ 11 \\ \hline 112 \end{array}$$

1. $\begin{array}{r} 123 \\ -\ 11 \\ \hline \end{array}$	2. $\begin{array}{r} 443 \\ -\ 77 \\ \hline \end{array}$	3. $\begin{array}{r} 318 \\ -\ 59 \\ \hline \end{array}$	4. $\begin{array}{r} 433 \\ -\ 78 \\ \hline \end{array}$	5. $\begin{array}{r} 311 \\ -\ 93 \\ \hline \end{array}$
6. $\begin{array}{r} 511 \\ -\ 66 \\ \hline \end{array}$	7. $\begin{array}{r} 333 \\ -\ 84 \\ \hline \end{array}$	8. $\begin{array}{r} 412 \\ -\ 55 \\ \hline \end{array}$	9. $\begin{array}{r} 233 \\ -\ 86 \\ \hline \end{array}$	10. $\begin{array}{r} 504 \\ -\ 37 \\ \hline \end{array}$
11. $\begin{array}{r} 711 \\ -\ 93 \\ \hline \end{array}$	12. $\begin{array}{r} 640 \\ -\ 77 \\ \hline \end{array}$	13. $\begin{array}{r} 411 \\ -\ 56 \\ \hline \end{array}$	14. $\begin{array}{r} 533 \\ -\ 76 \\ \hline \end{array}$	15. $\begin{array}{r} 711 \\ -\ 34 \\ \hline \end{array}$

LESSON 6 REVIEW A +
3 DIGIT + 2 DIGIT ADDITION & SUBTRACTION

$$\begin{array}{r} 505 \\ +\ 26 \\ \hline 531 \end{array}$$

1. $\begin{array}{r} 548 \\ +\ 19 \\ \hline \end{array}$	2. $\begin{array}{r} 6 \\ +\ 8 \\ \hline \end{array}$	3. $\begin{array}{r} 13 \\ +\ 5 \\ \hline \end{array}$	4. $\begin{array}{r} 146 \\ +\ 13 \\ \hline \end{array}$	5. $\begin{array}{r} 44 \\ +\ 76 \\ \hline \end{array}$
6. $\begin{array}{r} 453 \\ +\ 17 \\ \hline \end{array}$	7. $\begin{array}{r} 33 \\ +\ 8 \\ \hline \end{array}$	8. $\begin{array}{r} 146 \\ +\ 9 \\ \hline \end{array}$	9. $\begin{array}{r} 5006 \\ +\ 9 \\ \hline \end{array}$	10. $\begin{array}{r} 76 \\ +\ 13 \\ \hline \end{array}$
11. $\begin{array}{r} 14 \\ +\ 74 \\ \hline \end{array}$	12. $\begin{array}{r} 1977 \\ +\ 5 \\ \hline \end{array}$	13. $\begin{array}{r} 148 \\ +\ 44 \\ \hline \end{array}$	14. $\begin{array}{r} 889 \\ +\ 20 \\ \hline \end{array}$	15. $\begin{array}{r} 997 \\ +\ 15 \\ \hline \end{array}$

LESSON 6 REVIEW A –
3 DIGIT + 2 DIGIT ADDITION & SUBTRACTION

$$\begin{array}{r} 612 \\ -\ 48 \\ \hline 584 \end{array}$$

1. 443 − 88	2. 511 − 58	3. 407 − 38	4. 511 − 66	5. 611 − 83
6. 708 − 59	7. 453 − 77	8. 399 − 43	9. 716 − 58	10. 506 − 37
11. 613 − 77	12. 413 − 78	13. 603 − 54	14. 513 − 77	15. 603 − 88

LESSON 6 REVIEW B +
3 DIGIT + 2 DIGIT ADDITION & SUBTRACTION

$$\begin{array}{r} 628 \\ +\ 11 \\ \hline 639 \end{array}$$

1. $\begin{array}{r} 779 \\ +\ 39 \\ \hline \end{array}$	2. $\begin{array}{r} 481 \\ +\ 33 \\ \hline \end{array}$	3. $\begin{array}{r} 14 \\ +\ 74 \\ \hline \end{array}$	4. $\begin{array}{r} 609 \\ +\ 13 \\ \hline \end{array}$	5. $\begin{array}{r} 111 \\ +\ 76 \\ \hline \end{array}$
6. $\begin{array}{r} 876 \\ +\ 93 \\ \hline \end{array}$	7. $\begin{array}{r} 984 \\ +\ 78 \\ \hline \end{array}$	8. $\begin{array}{r} 769 \\ +\ 88 \\ \hline \end{array}$	9. $\begin{array}{r} 996 \\ +\ 99 \\ \hline \end{array}$	10. $\begin{array}{r} 980 \\ +\ 14 \\ \hline \end{array}$
11. $\begin{array}{r} 49 \\ +\ 79 \\ \hline \end{array}$	12. $\begin{array}{r} 763 \\ +\ 68 \\ \hline \end{array}$	13. $\begin{array}{r} 5 \\ +\ 9 \\ \hline \end{array}$	14. $\begin{array}{r} 19 \\ +\ 5 \\ \hline \end{array}$	15. $\begin{array}{r} 7693 \\ +\ 15 \\ \hline \end{array}$

LESSON 6 REVIEW B –
3 DIGIT + 2 DIGIT ADDITION & SUBTRACTION

$$\begin{array}{r} 345 \\ -\ 13 \\ \hline 332 \end{array}$$

1. 543 − 77	2. 613 − 85	3. 402 − 55	4. 419 − 33	5. 601 − 76
6. 408 − 59	7. 291 − 55	8. 763 − 555	9. 543 − 58	10. 605 − 38
11. 401 − 76	12. 333 − 64	13. 501 − 18	14. 502 − 19	15. 301 − 54

3/2 DIGIT SPEED TEST

1. 449 + 37	2. 217 + 48	3. 406 + 33	4. 947 + 13	5. 506 + 34
6. 140 - 28	7. 370 - 47	8. 504 - 68	9. 112 - 38	10. 607 - 49
11. 170 + 85	12. 490 + 93	13. 120 - 38	14. 300 - 47	15. 207 - 69
16. 500 - 88	17. 401 - 28	18. 870 - 26	19. 405 + 99	20. 270 - 46

LESSON 7 +
4 DIGIT + 2 DIGIT ADDITION & SUBTRACTION

$$\begin{array}{r} 1234 \\ +\quad 12 \\ \hline 1246 \end{array}$$

1. $\begin{array}{r} 4783 \\ +\ 14 \\ \hline \end{array}$	2. $\begin{array}{r} 5299 \\ +\ 23 \\ \hline \end{array}$	3. $\begin{array}{r} 7887 \\ +\ 56 \\ \hline \end{array}$	4. $\begin{array}{r} 9921 \\ +\ 44 \\ \hline \end{array}$	5. $\begin{array}{r} 7876 \\ +\ 56 \\ \hline \end{array}$
6. $\begin{array}{r} 8989 \\ +\ 99 \\ \hline \end{array}$	7. $\begin{array}{r} 6838 \\ +\ 78 \\ \hline \end{array}$	8. $\begin{array}{r} 9256 \\ +\ 73 \\ \hline \end{array}$	9. $\begin{array}{r} 4709 \\ +\ 95 \\ \hline \end{array}$	10. $\begin{array}{r} 7635 \\ +\ 53 \\ \hline \end{array}$
11. $\begin{array}{r} 1787 \\ +\ 99 \\ \hline \end{array}$	12. $\begin{array}{r} 5448 \\ +\ 97 \\ \hline \end{array}$	13. $\begin{array}{r} 4897 \\ +\ 87 \\ \hline \end{array}$	14. $\begin{array}{r} 5254 \\ +\ 94 \\ \hline \end{array}$	15. $\begin{array}{r} 7615 \\ +\ 88 \\ \hline \end{array}$

LESSON 7 -
4 DIGIT + 2 DIGIT ADDITION & SUBTRACTION

$$\begin{array}{r} 1234 \\ -\ \ 12 \\ \hline 1222 \end{array}$$

1. 8973 − 48	2. 5473 − 77	3. 4985 − 54	4. 5686 − 37	5. 7633 − 48
6. 7727 − 39	7. 4876 − 54	8. 4973 − 89	9. 9988 − 69	10. 1176 − 14
11. 5435 − 17	12. 7787 − 93	13. 4895 − 29	14. 5874 − 99	15. 7687 − 99

LESSON 7 REVIEW A +
4 DIGIT + 2 DIGIT ADDITION & SUBTRACTION

$$\begin{array}{r} 6000 \\ +\ \ 10 \\ \hline 6010 \end{array}$$

1. $\begin{array}{r} 7 \\ +\ 6 \\ \hline \end{array}$	2. $\begin{array}{r} 19 \\ +\ 36 \\ \hline \end{array}$	3. $\begin{array}{r} 33 \\ +\ 7 \\ \hline \end{array}$	4. $\begin{array}{r} 538 \\ +\ 26 \\ \hline \end{array}$	5. $\begin{array}{r} 1787 \\ +\ 38 \\ \hline \end{array}$
6. $\begin{array}{r} 156 \\ +\ 76 \\ \hline \end{array}$	7. $\begin{array}{r} 784 \\ +\ 19 \\ \hline \end{array}$	8. $\begin{array}{r} 5498 \\ +\ 78 \\ \hline \end{array}$	9. $\begin{array}{r} 3768 \\ +\ 59 \\ \hline \end{array}$	10. $\begin{array}{r} 17 \\ +\ 28 \\ \hline \end{array}$
11. $\begin{array}{r} 487 \\ +\ 67 \\ \hline \end{array}$	12. $\begin{array}{r} 331 \\ +\ 19 \\ \hline \end{array}$	13. $\begin{array}{r} 788 \\ +\ 54 \\ \hline \end{array}$	14. $\begin{array}{r} 5678 \\ +\ 93 \\ \hline \end{array}$	15. $\begin{array}{r} 7889 \\ +\ 89 \\ \hline \end{array}$

LESSON 7 REVIEW A –
4 DIGIT + 2 DIGIT ADDITION & SUBTRACTION

$$\begin{array}{r} 640 \\ -\ 20 \\ \hline 620 \end{array}$$

1. 1919 - 3	2. 2206 - 8	3. 1313 - 5	4. 4091 - 47	5. 6031 - 8
6. 4113 - 57	7. 6112 - 54	8. 8832 - 55	9. 7103 - 44	10. 3316 - 59
11. 2207 - 58	12. 3391 - 59	13. 1313 - 35	14. 1405 - 97	15. 1212 - 76

LESSON 7 REVIEW B +
4 DIGIT + 2 DIGIT ADDITION & SUBTRACTION

$$\begin{array}{r} 2225 \\ +\ \ 46 \\ \hline 2271 \end{array}$$

1. 4978 + 76	2. 7689 + 54	3. 4769 + 48	4. 5397 + 78	5. 9837 + 76
6. 7779 + 31	7. 458 + 97	8. 9789 + 36	9. 47 + 38	10. 17 + 6
11. 487 + 76	12. 5493 + 78	13. 4789 + 58	14. 487 + 76	15. 3893 + 17

LESSON 7 REVIEW B -
4 DIGIT + 2 DIGIT ADDITION & SUBTRACTION

$$\begin{array}{r} 3315 \\ -\ 26 \\ \hline 3279 \end{array}$$

1. $\begin{array}{r} 4403 \\ -\ 96 \\ \hline \end{array}$	2. $\begin{array}{r} 7604 \\ -\ 77 \\ \hline \end{array}$	3. $\begin{array}{r} 8811 \\ -\ 55 \\ \hline \end{array}$	4. $\begin{array}{r} 6703 \\ -\ 44 \\ \hline \end{array}$	5. $\begin{array}{r} 6111 \\ -\ 58 \\ \hline \end{array}$
6. $\begin{array}{r} 3386 \\ -\ 97 \\ \hline \end{array}$	7. $\begin{array}{r} 4407 \\ -\ 58 \\ \hline \end{array}$	8. $\begin{array}{r} 1100 \\ -\ 78 \\ \hline \end{array}$	9. $\begin{array}{r} 2101 \\ -\ 55 \\ \hline \end{array}$	10. $\begin{array}{r} 6101 \\ -\ 55 \\ \hline \end{array}$
11. $\begin{array}{r} 3407 \\ -\ 88 \\ \hline \end{array}$	12. $\begin{array}{r} 1703 \\ -\ 59 \\ \hline \end{array}$	13. $\begin{array}{r} 1403 \\ -\ 77 \\ \hline \end{array}$	14. $\begin{array}{r} 5113 \\ -\ 55 \\ \hline \end{array}$	15. $\begin{array}{r} 7683 \\ -\ 997 \\ \hline \end{array}$

4/2 DIGIT SPEED TEST

1. 4849 + 27	2. 2287 + 40	3. 4036 + 83	4. 9047 + 93	5. 5012 + 46
6. 3500 - 96	7. 3200 - 87	8. 9000 - 44	9. 1002 - 77	10. 4000 - 52
11. 2000 + 10	12. 9191 + 98	13. 1200 - 26	14. 7000 - 12	15. 1001 - 56
16. 5010 - 67	17. 1003 - 17	18. 6000 - 14	19. 7662 + 99	20. 1400 - 32

LESSON 8 +

3 DIGIT + 3 DIGIT ADDITION & SUBTRACTION

$$\begin{array}{r} 123 \\ +\ 123 \\ \hline 246 \end{array}$$

1. $\begin{array}{r} 128 \\ +\ 436 \\ \hline \end{array}$	2. $\begin{array}{r} 704 \\ +\ 192 \\ \hline \end{array}$	3. $\begin{array}{r} 768 \\ +\ 337 \\ \hline \end{array}$	4. $\begin{array}{r} 540 \\ +\ 196 \\ \hline \end{array}$	5. $\begin{array}{r} 670 \\ +\ 130 \\ \hline \end{array}$
6. $\begin{array}{r} 198 \\ +\ 337 \\ \hline \end{array}$	7. $\begin{array}{r} 456 \\ +\ 330 \\ \hline \end{array}$	8. $\begin{array}{r} 459 \\ +\ 137 \\ \hline \end{array}$	9. $\begin{array}{r} 477 \\ +\ 123 \\ \hline \end{array}$	10. $\begin{array}{r} 338 \\ +\ 476 \\ \hline \end{array}$
11. $\begin{array}{r} 456 \\ +\ 399 \\ \hline \end{array}$	12. $\begin{array}{r} 1073 \\ +\ 14 \\ \hline \end{array}$	13. $\begin{array}{r} 980 \\ +\ 450 \\ \hline \end{array}$	14. $\begin{array}{r} 393 \\ +\ 122 \\ \hline \end{array}$	15. $\begin{array}{r} 127 \\ +\ 450 \\ \hline \end{array}$

LESSON 8 -
3 DIGIT + 3 DIGIT ADDITION & SUBTRACTION

$$\begin{array}{r} 128 \\ -\ 123 \\ \hline 5 \end{array}$$

1. 603 − 156	2. 333 − 278	3. 413 − 193	4. 413 − 276	5. 501 − 399
6. 708 − 633	7. 592 − 586	8. 403 − 399	9. 768 − 539	10. 766 − 499
11. 408 − 309	12. 766 − 599	13. 806 − 793	14. 544 − 399	15. 676 − 598

LESSON 8 REVIEW A –
3 DIGIT + 3 DIGIT ADDITION & SUBTRACTION

$$\begin{array}{r} 600 \\ -\ 130 \\ \hline 470 \end{array}$$

1. 609 - 534	2. 463 - 299	3. 706 - 488	4. 902 - 788	5. 691 - 566
6. 444 - 299	7. 966 - 899	8. 888 - 399	9. 204 - 156	10. 773 - 699
11. 302 - 288	12. 706 - 498	13. 418 - 287	14. 332 - 276	15. 296 - 188

LESSON 8 REVIEW A +
3 DIGIT + 3 DIGIT ADDITION & SUBTRACTION

505
+ 26
531

1. 4503 + 6	2. 453 + 17	3. 701 + 53	4. 999 + 999	5. 6 + 8
6. 1700 + 30	7. 16 + 9	8. 23 + 40	9. 17 + 9	10. 459 + 459
11. 109 + 78	12. 70 + 3	13. 4993 + 12	14. 576 + 89	15. 936 + 72

LESSON 8 REVIEW B -
3 DIGIT + 3 DIGIT ADDITION & SUBTRACTION

$$\begin{array}{r} 711 \\ -\ 611 \\ \hline 100 \end{array}$$

1. $\begin{array}{r} 602 \\ -\ 198 \\ \hline \end{array}$	2. $\begin{array}{r} 544 \\ -\ 299 \\ \hline \end{array}$	3. $\begin{array}{r} 278 \\ -\ 199 \\ \hline \end{array}$	4. $\begin{array}{r} 504 \\ -\ 377 \\ \hline \end{array}$	5. $\begin{array}{r} 612 \\ -\ 338 \\ \hline \end{array}$
6. $\begin{array}{r} 444 \\ -\ 398 \\ \hline \end{array}$	7. $\begin{array}{r} 767 \\ -\ 499 \\ \hline \end{array}$	8. $\begin{array}{r} 786 \\ -\ 397 \\ \hline \end{array}$	9. $\begin{array}{r} 403 \\ -\ 207 \\ \hline \end{array}$	10. $\begin{array}{r} 202 \\ -\ 157 \\ \hline \end{array}$
11. $\begin{array}{r} 446 \\ -\ 297 \\ \hline \end{array}$	12. $\begin{array}{r} 288 \\ -\ 179 \\ \hline \end{array}$	13. $\begin{array}{r} 442 \\ -\ 388 \\ \hline \end{array}$	14. $\begin{array}{r} 332 \\ -\ 263 \\ \hline \end{array}$	15. $\begin{array}{r} 602 \\ -\ 573 \\ \hline \end{array}$

LESSON 8 REVIEW B +
3 DIGIT + 3 DIGIT ADDITION & SUBTRACTION

$$\begin{array}{r} 123 \\ +\,321 \\ \hline 444 \end{array}$$

1. $\begin{array}{r} 2700 \\ +\quad 14 \\ \hline \end{array}$	2. $\begin{array}{r} 2304 \\ +\quad 53 \\ \hline \end{array}$	3. $\begin{array}{r} 698 \\ +\quad 53 \\ \hline \end{array}$	4. $\begin{array}{r} 1809 \\ +\quad 58 \\ \hline \end{array}$	5. $\begin{array}{r} 970 \\ +\ 136 \\ \hline \end{array}$
6. $\begin{array}{r} 983 \\ +\ 128 \\ \hline \end{array}$	7. $\begin{array}{r} 4590 \\ +\ 1376 \\ \hline \end{array}$	8. $\begin{array}{r} 765 \\ +\ 333 \\ \hline \end{array}$	9. $\begin{array}{r} 177 \\ +\,589 \\ \hline \end{array}$	10. $\begin{array}{r} 873 \\ +\ 234 \\ \hline \end{array}$
11. $\begin{array}{r} 6789 \\ +\quad 43 \\ \hline \end{array}$	12. $\begin{array}{r} 5413 \\ +\quad 28 \\ \hline \end{array}$	13. $\begin{array}{r} 7631 \\ +\quad 58 \\ \hline \end{array}$	14. $\begin{array}{r} 673 \\ +\quad 53 \\ \hline \end{array}$	15. $\begin{array}{r} 834 \\ +\,129 \\ \hline \end{array}$

3/3 DIGIT SPEED TEST

1. 749 + 604	2. 314 + 266	3. 457 +196	4. 336 + 714	5. 706 + 246
6. 300 - 196	7. 820 - 187	8. 903 - 264	9. 702 - 577	10. 400 - 176
11. 400 +199	12. 415 + 234	13. 400 - 334	14. 200 - 176	15. 100 - 256
16. 600 - 487	17. 210 - 136	18. 400 -118	19. 140 + 266	20. 340 - 312

LESSON 9 +
4 DIGIT + 3 DIGIT ADDITION & SUBTRACTION

$$\begin{array}{r} 1234 \\ +\ 123 \\ \hline 1357 \end{array}$$

1. 2076 + 111	2. 2487 + 503	3. 4607 + 309	4. 6700 + 433	5. 5046 + 372
6. 4408 + 336	7. 5409 + 330	8. 4607 + 223	9. 6703 + 144	10. 5307 + 690
11. 8931 + 407	12. 6817 + 301	13. 4703 + 112	14. 9779 + 311	15. 2678 + 443

LESSON 9 -
4 DIGIT + 3 DIGIT ADDITION & SUBTRACTION

$$\begin{array}{r} 3702 \\ -\ 123 \\ \hline 3579 \end{array}$$

1. 1114 − 276	2. 2206 − 479	3. 4437 − 998	4. 2203 − 776	5. 3313 − 547
6. 4397 − 888	7. 1145 − 377	8. 2031 − 598	9. 1454 − 766	10. 5042 − 777
11. 4407 − 593	12. 7061 − 887	13. 1407 − 998	14. 5402 − 399	15. 1691 − 555

LESSON 9 REVIEW A +
4 DIGIT + 3 DIGIT ADDITION & SUBTRACTION

$$\begin{array}{r} 5555 \\ +\ 13 \\ \hline \end{array}$$

1. $\begin{array}{r} 7 \\ +\ 6 \\ \hline \end{array}$	2. $\begin{array}{r} 13 \\ +\ 58 \\ \hline \end{array}$	3. $\begin{array}{r} 703 \\ +\ 14 \\ \hline \end{array}$	4. $\begin{array}{r} 1238 \\ +\ 14 \\ \hline \end{array}$	5. $\begin{array}{r} 6708 \\ +\ 376 \\ \hline \end{array}$
6. $\begin{array}{r} 14 \\ +\ 33 \\ \hline \end{array}$	7. $\begin{array}{r} 148 \\ +\ 53 \\ \hline \end{array}$	8. $\begin{array}{r} 6783 \\ +\ 409 \\ \hline \end{array}$	9. $\begin{array}{r} 7633 \\ +\ 408 \\ \hline \end{array}$	10. $\begin{array}{r} 1733 \\ +\ 533 \\ \hline \end{array}$
11. $\begin{array}{r} 5431 \\ +\ 376 \\ \hline \end{array}$	12. $\begin{array}{r} 7787 \\ +\ 936 \\ \hline \end{array}$	13. $\begin{array}{r} 1099 \\ +\ 999 \\ \hline \end{array}$	14. $\begin{array}{r} 8888 \\ +\ 888 \\ \hline \end{array}$	15. $\begin{array}{r} 763 \\ +\ 14 \\ \hline \end{array}$

LESSON 9 REVIEW A -
4 DIGIT + 3 DIGIT ADDITION & SUBTRACTION

$$\begin{array}{r} 4439 \\ +\quad 11 \\ \hline 4450 \end{array}$$

1. $\begin{array}{r} 5504 \\ -786 \\ \hline \end{array}$	2. $\begin{array}{r} 4312 \\ -\ 554 \\ \hline \end{array}$	3. $\begin{array}{r} 6711 \\ -\ 333 \\ \hline \end{array}$	4. $\begin{array}{r} 1419 \\ -\ 786 \\ \hline \end{array}$	5. $\begin{array}{r} 5405 \\ -\ 378 \\ \hline \end{array}$
6. $\begin{array}{r} 1111 \\ -\ 447 \\ \hline \end{array}$	7. $\begin{array}{r} 3333 \\ -\ 645 \\ \hline \end{array}$	8. $\begin{array}{r} 2221 \\ -\ 776 \\ \hline \end{array}$	9. $\begin{array}{r} 4405 \\ -\ 788 \\ \hline \end{array}$	10. $\begin{array}{r} 9999 \\ -\ 397 \\ \hline \end{array}$
11. $\begin{array}{r} 7800 \\ -\ 764 \\ \hline \end{array}$	12. $\begin{array}{r} 5400 \\ -\ 778 \\ \hline \end{array}$	13. $\begin{array}{r} 1313 \\ -\ 444 \\ \hline \end{array}$	14. $\begin{array}{r} 4071 \\ -\ 399 \\ \hline \end{array}$	15. $\begin{array}{r} 6101 \\ -\ 888 \\ \hline \end{array}$

LESSON 9 REVIEW B +
4 DIGIT + 3 DIGIT ADDITION & SUBTRACTION

$$\begin{array}{r} 2234 \\ +\ \ 26 \\ \hline \end{array}$$

1. $\begin{array}{r} 19 \\ +\ 3 \\ \hline \end{array}$	2. $\begin{array}{r} 1787 \\ +\ 787 \\ \hline \end{array}$	3. $\begin{array}{r} 3498 \\ +\ 459 \\ \hline \end{array}$	4. $\begin{array}{r} 786 \\ +\ 153 \\ \hline \end{array}$	5. $\begin{array}{r} 678 \\ +\ 988 \\ \hline \end{array}$
6. $\begin{array}{r} 6853 \\ +\ 553 \\ \hline \end{array}$	7. $\begin{array}{r} 2806 \\ +\ 311 \\ \hline \end{array}$	8. $\begin{array}{r} 5408 \\ +\ 533 \\ \hline \end{array}$	9. $\begin{array}{r} 678 \\ +\ 133 \\ \hline \end{array}$	10. $\begin{array}{r} 4831 \\ +\ 193 \\ \hline \end{array}$
11. $\begin{array}{r} 9763 \\ +\ 889 \\ \hline \end{array}$	12. $\begin{array}{r} 5391 \\ +\ 196 \\ \hline \end{array}$	13. $\begin{array}{r} 6309 \\ +\ 1406 \\ \hline \end{array}$	14. $\begin{array}{r} 590 \\ +\ 76 \\ \hline \end{array}$	15. $\begin{array}{r} 6 \\ +\ 5 \\ \hline \end{array}$

LESSON 9 REVIEW B –
4 DIGIT + 3 DIGIT ADDITION & SUBTRACTION

$$\begin{array}{r} 1106 \\ +\ 110 \\ \hline 1216 \end{array}$$

1. 1212 − 543	2. 3342 − 777	3. 4402 − 789	4. 1403 − 976	5. 1616 − 539
6. 4071 − 587	7. 1401 − 583	8. 7701 − 539	9. 7070 − 556	10. 1110 − 789
11. 4091 − 793	12. 4003 − 566	13. 1411 − 776	14. 5003 − 459	15. 6113 − 447

4/3 DIGIT SPEED TEST

1. 7249 + 404	2. 5814 + 346	3. 8667 + 276	4. 4506 - 384	5. 8700 + 448
6. 4700 - 266	7. 2786 - 487	8. 9242 - 864	9. 7360 - 490	10. 2600 - 456
11. 4036 + 149	12. 8705 + 454	13. 2600 - 474	14. 3800 - 119	15. 2600 - 446
16. 6400 - 397	17. 2610 - 486	18. 3400 - 117	19. 4940 + 706	20. 1940 - 707

LESSON 10 +
4 DIGIT + 4 DIGIT ADDITION & SUBTRACTION

$$\begin{array}{r} 1234 \\ +\,1234 \\ \hline 2468 \end{array}$$

1. $\begin{array}{r} 4003 \\ +\ \ 1673 \\ \hline \end{array}$	2. $\begin{array}{r} 4309 \\ +\ \ 3333 \\ \hline \end{array}$	3. $\begin{array}{r} 1309 \\ +\ \ 1276 \\ \hline \end{array}$	4. $\begin{array}{r} 1205 \\ +\,3309 \\ \hline \end{array}$	5. $\begin{array}{r} 6703 \\ +\,1143 \\ \hline \end{array}$
6. $\begin{array}{r} 5793 \\ +\ \ 1608 \\ \hline \end{array}$	7. $\begin{array}{r} 8903 \\ +\ \ 1766 \\ \hline \end{array}$	8. $\begin{array}{r} 1437 \\ +\ \ 5203 \\ \hline \end{array}$	9. $\begin{array}{r} 1733 \\ +\ \ 1647 \\ \hline \end{array}$	10. $\begin{array}{r} 3803 \\ +\ \ 2717 \\ \hline \end{array}$
11. $\begin{array}{r} 9307 \\ +\ \ 1449 \\ \hline \end{array}$	12. $\begin{array}{r} 7639 \\ +\ \ 4205 \\ \hline \end{array}$	13. $\begin{array}{r} 7863 \\ +\ \ 1986 \\ \hline \end{array}$	14. $\begin{array}{r} 7779 \\ +\ \ 3110 \\ \hline \end{array}$	15. $\begin{array}{r} 6783 \\ +\ \ 7789 \\ \hline \end{array}$

LESSON 10 -
4 DIGIT + 4 DIGIT ADDITION & SUBTRACTION

$$\begin{array}{r} 6110 \\ -\ 1500 \\ \hline 4610 \end{array}$$

1. 1234 − 1214	2. 4093 − 2777	3. 5116 − 3887	4. 1163 − 1077	5. 1111 − 1099
6. 3401 − 2776	7. 4803 − 3977	8. 8764 − 5988	9. 4708 − 3999	10. 1076 − 1088
11. 4308 − 3679	12. 5367 − 4498	13. 4333 − 3778	14. 5092 − 3883	15. 6113 − 5437

LESSON 10 SUB REVIEW A +
4 DIGIT + 4 DIGIT ADDITION & SUBTRACTION

$$\begin{array}{r} 6110 \\ +\ 2250 \\ \hline 8360 \end{array}$$

1. $\begin{array}{r} 4312 \\ -\ 2793 \\ \hline \end{array}$	2. $\begin{array}{r} 4117 \\ -\ 2988 \\ \hline \end{array}$	3. $\begin{array}{r} 4111 \\ -\ 3886 \\ \hline \end{array}$	4. $\begin{array}{r} 2207 \\ -\ 1788 \\ \hline \end{array}$	5. $\begin{array}{r} 5116 \\ -\ 2778 \\ \hline \end{array}$
6. $\begin{array}{r} 7113 \\ -\ 5664 \\ \hline \end{array}$	7. $\begin{array}{r} 8103 \\ -\ 6977 \\ \hline \end{array}$	8. $\begin{array}{r} 7609 \\ -\ 4877 \\ \hline \end{array}$	9. $\begin{array}{r} 4332 \\ -\ 2994 \\ \hline \end{array}$	10. $\begin{array}{r} 7061 \\ -\ 3884 \\ \hline \end{array}$
11. $\begin{array}{r} 1133 \\ -\ 1107 \\ \hline \end{array}$	12. $\begin{array}{r} 4511 \\ -\ 3997 \\ \hline \end{array}$	13. $\begin{array}{r} 4141 \\ -\ 3989 \\ \hline \end{array}$	14. $\begin{array}{r} 7116 \\ -\ 5338 \\ \hline \end{array}$	15. $\begin{array}{r} 8102 \\ -\ 5788 \\ \hline \end{array}$

LESSON 10 REVIEW B -
4 DIGIT + 4 DIGIT ADDITION & SUBTRACTION

$$\begin{array}{r} 1025 \\ +\ 1300 \\ \hline 275 \end{array}$$

1. 2237 − 1449	2. 1003 − 984	3. 4516 − 3787	4. 5113 − 4876	5. 7613 − 5877
6. 3313 − 2784	7. 5446 − 3788	8. 4076 − 2977	9. 3603 − 2876	10. 3331 − 2762
11. 4653 − 3784	12. 7604 − 5995	13. 8324 − 6545	14. 2677 − 2588	15. 3832 − 2944

LESSON 10 REVIEW A +
4 DIGIT + 4 DIGIT ADDITION & SUBTRACTION

3111
+ 2154
6265

1. 5043 + 3076	2. 7609 + 9322	3. 1789 + 3309	4. 6709 + 3309	5. 3391 + 1205
6. 7639 + 1409	7. 1337 + 4058	8. 4376 + 2273	9. 4599 + 3366	10. 4439 + 17
11. 17 + 58	12. 339 + 76	13. 1405 + 126	14. 1718 + 533	15. 14 + 9

LESSON 10 REVIEW B +
4 DIGIT + 4 DIGIT ADDITION & SUBTRACTION

$$\begin{array}{r} 4256 \\ +\ 8709 \\ \hline 12{,}965 \end{array}$$

1. $\begin{array}{r} 3 \\ +\ 3 \\ \hline \end{array}$	2. $\begin{array}{r} 6 \\ +\ 6 \\ \hline \end{array}$	3. $\begin{array}{r} 9 \\ +\ 9 \\ \hline \end{array}$	4. $\begin{array}{r} 12 \\ +\ 12 \\ \hline \end{array}$	5. $\begin{array}{r} 15 \\ +\ 15 \\ \hline \end{array}$
6. $\begin{array}{r} 18 \\ +\ 18 \\ \hline \end{array}$	7. $\begin{array}{r} 21 \\ +\ 21 \\ \hline \end{array}$	8. $\begin{array}{r} 24 \\ +\ 24 \\ \hline \end{array}$	9. $\begin{array}{r} 27 \\ +\ 27 \\ \hline \end{array}$	10. $\begin{array}{r} 30 \\ +\ 30 \\ \hline \end{array}$
11. $\begin{array}{r} 333 \\ +\ 333 \\ \hline \end{array}$	12. $\begin{array}{r} 3336 \\ +\ 3636 \\ \hline \end{array}$	13. $\begin{array}{r} 9797 \\ +\ 9999 \\ \hline \end{array}$	14. $\begin{array}{r} 8889 \\ +\ 9899 \\ \hline \end{array}$	15. $\begin{array}{r} 7777 \\ +\ 6893 \\ \hline \end{array}$

Bonus 16. $\begin{array}{r} 58{,}766 \\ +\ 69{,}938 \\ \hline \end{array}$	**Bonus** 17. $\begin{array}{r} 999{,}983 \\ +\ 76{,}437 \\ \hline \end{array}$

4/4 DIGIT SPEED TEST

1. 7249 + 2604	2. 4414 + 2746	3. 2967 + 4876	4. 4384 - 4384	5. 1400 + 3848
6. 4700 - 1766	7. 3686 - 1687	8. 1442 - 1164	9. 7399 - 4790	10. 2700 - 1456
11. 3336 + 2649	12. 7005 + 7054	13. 5400 - 4774	14. 9700 - 3719	15. 1300 - 1046
16. 4900 - 2797	17. 7710 - 6686	18. 5600 -3417	19. 2740 + 1306	20. 4740 - 1007

Multiplication & Division

This section introduces you to an easier quicker way to add and subtract.

Multiplication is represented by these symbols: x, *, .

MULTIPLICATION

The fastest way to add is by multiplying. Multiplication makes addition quicker by putting things into groups. For instance, Dr. Jones wants to know how many assignments she needs to grade this weekend. Three students each turned in two assignments. How many assignments in all will she have to grade?

There are three groups of two. To solve this problem, we could count each assignment or we could memorize this grouping using whats called mutiplication facts. $3 \times 2 = 6$ *or* $2 * 3 = 6$ *or* $2(3) = 6$

Multiplication & Division

This section introduces you to an easier quicker way to add and subtract.

Division is represented by these symbols: /, ⟌, ÷

DIVISION

Division is the opposite of multiplication. If you look closely at the answers to your multiplication facts, you will see that they are just backwards division. It's really the same relationship that subtraction has with addition. The fast way to subtract is division. Division makes subtraction quicker by splitting things up. For instance, Dr. Jones wants to know how many of the assignments she is going to grade this weekend are from Ana.

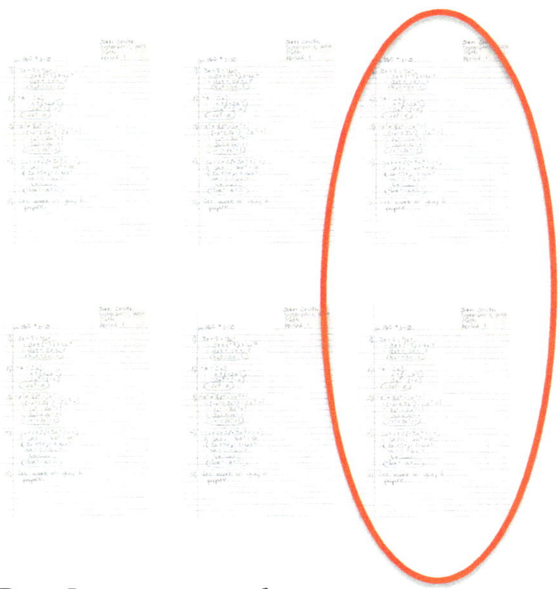

Dr. Jones sorts the papers into two groups of three to see which ones are Ana's. To solve this problem, we could sort the papers and arrange them according to each student to see how many are from Ana. Or we could do the opposite of multiplication and break our group of six into three rows of two.

$$6 \div 3 = 2 \quad 6 \div 2 = 3$$

L11 x
1 Digit x 1 Digit Multiplication & Division

$$\begin{array}{r} 8 \\ \times\ 3 \\ \hline 24 \end{array}$$

1. $\begin{array}{r} 5 \\ \times\ 4 \\ \hline \end{array}$	2. $\begin{array}{r} 3 \\ \times\ 7 \\ \hline \end{array}$	3. $\begin{array}{r} 6 \\ \times\ 2 \\ \hline \end{array}$	4. $\begin{array}{r} 5 \\ \times\ 8 \\ \hline \end{array}$	5. $\begin{array}{r} 6 \\ \times\ 3 \\ \hline \end{array}$
6. $\begin{array}{r} 7 \\ \times\ 3 \\ \hline \end{array}$	7. $\begin{array}{r} 8 \\ \times\ 4 \\ \hline \end{array}$	8. $\begin{array}{r} 9 \\ \times\ 6 \\ \hline \end{array}$	9. $\begin{array}{r} 7 \\ \times\ 5 \\ \hline \end{array}$	10. $\begin{array}{r} 2 \\ \times\ 2 \\ \hline \end{array}$
11. $\begin{array}{r} 4 \\ \times\ 6 \\ \hline \end{array}$	12. $\begin{array}{r} 5 \\ \times\ 6 \\ \hline \end{array}$	13. $\begin{array}{r} 4 \\ \times\ 9 \\ \hline \end{array}$	14. $\begin{array}{r} 3 \\ \times\ 1 \\ \hline \end{array}$	15. $\begin{array}{r} 0 \\ \times\ 7 \\ \hline \end{array}$

L11.5 ÷
1 Digit x 1 Digit Multiplication & Division

$$2\overline{)6}$$ = 3

1. $3\overline{)9}$	2. $2\overline{)8}$	3. $4\overline{)8}$	4. $2\overline{)6}$	5. $1\overline{)8}$
6. $2\overline{)14}$	7. $3\overline{)16}$	8. $8\overline{)8}$	9. $2\overline{)20}$	10. $2\overline{)12}$
11. $3\overline{)24}$	12. $2\overline{)18}$	13. $4\overline{)16}$	14. $2\overline{)10}$	15. $5\overline{)60}$

L11.5 REVIEW A *

1 Digit x 1 Digit Multiplication & Division

3
x 7

1. 2 x 2	2. 2 x 4	3. 2 x 5	4. 2 x 6	5. 2 x 7
6. 2 x 4	7. 3 x 3	8. 9 x 3	9. 5 x 3	10. 4 x 7
11. 1 x 4	12. 9 x 4	13. 9 x 5	14. 5 x 5	15. 6 x 5

L11 REVIEW B ÷

1 Digit x 1 Digit Multiplication & Division

2
4) 8

1. 8) 9	2. 7) 7	3. 2) 5	4. 3) 3	5. 5) 7
6. 9) 9	7. 6) 8	8. 5) 9	9. 3) 9	10. 6) 9
11. 7) 9	12. 3) 8	13. 3) 7	14. 4) 6	15. 5) 8

L11.5 REVIEW A x
1 Digit x 1 Digit Multiplication & Division

$$\begin{array}{r} 9 \\ \times\ 2 \\ \hline 18 \end{array}$$

1. $\begin{array}{r} 5 \\ \times\ 4 \\ \hline \end{array}$	2. $\begin{array}{r} 3 \\ \times\ 7 \\ \hline \end{array}$	3. $\begin{array}{r} 6 \\ \times\ 2 \\ \hline \end{array}$	4. $\begin{array}{r} 5 \\ \times\ 8 \\ \hline \end{array}$	5. $\begin{array}{r} 6 \\ \times\ 3 \\ \hline \end{array}$
6. $\begin{array}{r} 7 \\ \times\ 3 \\ \hline \end{array}$	7. $\begin{array}{r} 8 \\ \times\ 4 \\ \hline \end{array}$	8. $\begin{array}{r} 9 \\ \times\ 6 \\ \hline \end{array}$	9. $\begin{array}{r} 7 \\ \times\ 5 \\ \hline \end{array}$	10. $\begin{array}{r} 2 \\ \times\ 2 \\ \hline \end{array}$
11. $\begin{array}{r} 4 \\ \times\ 6 \\ \hline \end{array}$	12. $\begin{array}{r} 5 \\ \times\ 6 \\ \hline \end{array}$	13. $\begin{array}{r} 4 \\ \times\ 9 \\ \hline \end{array}$	14. $\begin{array}{r} 3 \\ \times\ 1 \\ \hline \end{array}$	15. $\begin{array}{r} 0 \\ \times\ 7 \\ \hline \end{array}$

L11.5 REVIEW B ÷
1 Digit x 1 Digit Multiplication & Division

$$\begin{array}{r} 2 \\ 2\overline{)4} \end{array}$$

1. $8\overline{)9}$	2. $7\overline{)7}$	3. $2\overline{)5}$	4. $3\overline{)3}$	5. $5\overline{)7}$
6. $9\overline{)9}$	7. $6\overline{)8}$	8. $5\overline{)9}$	9. $3\overline{)9}$	10. $6\overline{)9}$
11. $7\overline{)9}$	12. $3\overline{)8}$	13. $3\overline{)7}$	14. $4\overline{)6}$	15. $5\overline{)8}$

1/1 DIGIT SPEED TEST

1. $6\overline{)54}$	2. 3 x 4	3. 1 x 5	4. $3\overline{)3}$	5. 5 x 7
6. 5 x 3	7. $6\overline{)8}$	8. 7 x 1	9. 4 x 2	10. $6\overline{)9}$
11. $7\overline{)63}$	12. 8 x 2	13. 2 x 4	14. 6 x 9	15. $5\overline{)40}$
16. 2 x 1	17. 8 x 6	18. $3\overline{)7}$	19. 3 x 2	20. 4 x 4

L12 x
2 Digit x 1 Multiplication & Division

$$\begin{array}{r} 12 \\ \times\ 2 \\ \hline 24 \end{array}$$

1. $\begin{array}{r} 10 \\ \times\ 3 \\ \hline \end{array}$	2. $\begin{array}{r} 12 \\ \times\ 6 \\ \hline \end{array}$	3. $\begin{array}{r} 11 \\ \times\ 5 \\ \hline \end{array}$	4. $\begin{array}{r} 12 \\ \times\ 3 \\ \hline \end{array}$	5. $\begin{array}{r} 15 \\ \times\ 5 \\ \hline \end{array}$
6. $\begin{array}{r} 23 \\ \times\ 3 \\ \hline \end{array}$	7. $\begin{array}{r} 14 \\ \times\ 5 \\ \hline \end{array}$	8. $\begin{array}{r} 22 \\ \times\ 3 \\ \hline \end{array}$	9. $\begin{array}{r} 36 \\ \times\ 3 \\ \hline \end{array}$	10. $\begin{array}{r} 44 \\ \times\ 5 \\ \hline \end{array}$
11. $\begin{array}{r} 14 \\ \times\ 6 \\ \hline \end{array}$	12. $\begin{array}{r} 13 \\ \times\ 5 \\ \hline \end{array}$	13. $\begin{array}{r} 12 \\ \times\ 8 \\ \hline \end{array}$	14. $\begin{array}{r} 11 \\ \times\ 9 \\ \hline \end{array}$	15. $\begin{array}{r} 93 \\ \times\ 6 \\ \hline \end{array}$

L12 ÷
2 Digit x 1 Multiplication & Division

$$5\overline{)45} = 9$$

1. $9\overline{)45}$	2. $5\overline{)45}$	3. $3\overline{)24}$	4. $6\overline{)36}$	5. $8\overline{)64}$
6. $2\overline{)14}$	7. $3\overline{)16}$	8. $8\overline{)57}$	9. $2\overline{)63}$	10. $2\overline{)72}$
11. $3\overline{)27}$	12. $2\overline{)33}$	13. $4\overline{)44}$	14. $7\overline{)35}$	15. $6\overline{)86}$

L12.5 REVIEW A x
2 Digit x 1 Multiplication & Division

$$\begin{array}{r} 35 \\ \times\ 8 \\ \hline 280 \end{array}$$

1. $\begin{array}{r} 44 \\ \times\ 6 \\ \hline \end{array}$	2. $\begin{array}{r} 33 \\ \times\ 7 \\ \hline \end{array}$	3. $\begin{array}{r} 24 \\ \times\ 5 \\ \hline \end{array}$	4. $\begin{array}{r} 16 \\ \times\ 5 \\ \hline \end{array}$	5. $\begin{array}{r} 27 \\ \times\ 3 \\ \hline \end{array}$
6. $\begin{array}{r} 17 \\ \times\ 1 \\ \hline \end{array}$	7. $\begin{array}{r} 88 \\ \times\ 3 \\ \hline \end{array}$	8. $\begin{array}{r} 96 \\ \times\ 5 \\ \hline \end{array}$	9. $\begin{array}{r} 76 \\ \times\ 5 \\ \hline \end{array}$	10. $\begin{array}{r} 14 \\ \times\ 8 \\ \hline \end{array}$
11. $\begin{array}{r} 47 \\ \times\ 2 \\ \hline \end{array}$	12. $\begin{array}{r} 38 \\ \times\ 6 \\ \hline \end{array}$	13. $\begin{array}{r} 12 \\ \times\ 8 \\ \hline \end{array}$	14. $\begin{array}{r} 19 \\ \times\ 3 \\ \hline \end{array}$	15. $\begin{array}{r} 15 \\ \times\ 4 \\ \hline \end{array}$

L12 REVIEW A ÷
2 Digit x 1 Multiplication & Division

$$2\overline{)60}\quad 30$$

1. $9\overline{)36}$	2. $5\overline{)55}$	3. $3\overline{)60}$	4. $6\overline{)48}$	5. $8\overline{)96}$
6. $2\overline{)30}$	7. $3\overline{)33}$	8. $7\overline{)99}$	9. $6\overline{)64}$	10. $3\overline{)84}$
11. $8\overline{)28}$	12. $4\overline{)56}$	13. $5\overline{)65}$	14. $9\overline{)87}$	15. $5\overline{)90}$

L12 REVIEW B x
2 Digit x 1 Multiplication & Division

$$\begin{array}{r} 12 \\ \times\ 5 \\ \hline 60 \end{array}$$

1. $\begin{array}{r} 54 \\ \times\ 3 \\ \hline \end{array}$	2. $\begin{array}{r} 28 \\ \times\ 7 \\ \hline \end{array}$	3. $\begin{array}{r} 14 \\ \times\ 3 \\ \hline \end{array}$	4. $\begin{array}{r} 45 \\ \times\ 6 \\ \hline \end{array}$	5. $\begin{array}{r} 12 \\ \times\ 5 \\ \hline \end{array}$
6. $\begin{array}{r} 13 \\ \times\ 6 \\ \hline \end{array}$	7. $\begin{array}{r} 88 \\ \times\ 4 \\ \hline \end{array}$	8. $\begin{array}{r} 92 \\ \times\ 3 \\ \hline \end{array}$	9. $\begin{array}{r} 16 \\ \times\ 7 \\ \hline \end{array}$	10. $\begin{array}{r} 11 \\ \times\ 8 \\ \hline \end{array}$
11. $\begin{array}{r} 45 \\ \times\ 1 \\ \hline \end{array}$	12. $\begin{array}{r} 68 \\ \times\ 9 \\ \hline \end{array}$	13. $\begin{array}{r} 17 \\ \times\ 3 \\ \hline \end{array}$	14. $\begin{array}{r} 10 \\ \times\ 6 \\ \hline \end{array}$	15. $\begin{array}{r} 47 \\ \times\ 0 \\ \hline \end{array}$

L12 REVIEW B ÷
2 Digit x 1 Multiplication & Division

$$8\overline{)88}$$ = 11

1. $6\overline{)34}$	2. $4\overline{)64}$	3. $3\overline{)66}$	4. $7\overline{)46}$	5. $9\overline{)27}$
6. $4\overline{)84}$	7. $9\overline{)15}$	8. $6\overline{)19}$	9. $2\overline{)28}$	10. $5\overline{)44}$
11. $6\overline{)37}$	12. $7\overline{)37}$	13. $5\overline{)49}$	14. $8\overline{)37}$	15. $7\overline{)28}$

2/1 DIGIT SPEED TEST

1. $8\overline{)56}$	2. 40×4	3. 96×5	4. $9\overline{)72}$	5. 76×7
6. 38×3	7. $5\overline{)45}$	8. 88×3	9. 67×4	10. $3\overline{)44}$
11. $2\overline{)48}$	12. 58×2	13. 76×4	14. 64×7	15. $8\overline{)40}$
16. 43×4	17. 67×6	18. $8\overline{)48}$	19. 45×6	20. 77×4

L13 x
3 Digit x 1 Multiplication & Division

$$\begin{array}{r} 123 \\ \underline{X \quad 1} \\ 123 \end{array}$$

1. $\begin{array}{r} 154 \\ \underline{x \quad 2} \end{array}$	2. $\begin{array}{r} 130 \\ \underline{x \quad 5} \end{array}$	3. $\begin{array}{r} 140 \\ \underline{x \quad 3} \end{array}$	4. $\begin{array}{r} 158 \\ \underline{x \quad 0} \end{array}$	5. $\begin{array}{r} 122 \\ \underline{x \quad 2} \end{array}$
6. $\begin{array}{r} 712 \\ \underline{x \quad 5} \end{array}$	7. $\begin{array}{r} 833 \\ \underline{x \quad 3} \end{array}$	8. $\begin{array}{r} 915 \\ \underline{x \quad 5} \end{array}$	9. $\begin{array}{r} 962 \\ \underline{x \quad 6} \end{array}$	10. $\begin{array}{r} 102 \\ \underline{x \quad 3} \end{array}$
11. $\begin{array}{r} 414 \\ \underline{x \quad 4} \end{array}$	12. $\begin{array}{r} 306 \\ \underline{x \quad 5} \end{array}$	13. $\begin{array}{r} 133 \\ \underline{x \quad 6} \end{array}$	14. $\begin{array}{r} 154 \\ \underline{x \quad 9} \end{array}$	15. $\begin{array}{r} 106 \\ \underline{x \quad 8} \end{array}$

L13 ÷
3 Digit x 1 Multiplication & Division

$$1\overline{)123}$$ = 123

1. $5\overline{)550}$	2. $3\overline{)345}$	3. $2\overline{)660}$	4. $7\overline{)469}$	5. $9\overline{)279}$
6. $5\overline{)845}$	7. $8\overline{)158}$	8. $6\overline{)293}$	9. $7\overline{)280}$	10. $5\overline{)445}$
11. $3\overline{)370}$	12. $9\overline{)347}$	13. $5\overline{)459}$	14. $8\overline{)537}$	15. $7\overline{)208}$

L13 REVIEW A x
3 Digit x 1 Multiplication & Division

$$\begin{array}{r} 100 \\ \underline{\times \quad 6} \\ 600 \end{array}$$

1. 254 x 3	2. 356 x 2	3. 125 x 5	4. 500 x 4	5. 500 x 4
6. 955 x 9	7. 726 x 8	8. 555 x 5	9. 899 x 7	10. 936 x 7
11. 321 x 4	12. 150 x 8	13. 133 x 6	14. 446 x 3	15. 535 x 2

L13 REVIEW B ÷
3 Digit x 1 Multiplication & Division

$$5\overline{)519} = 103 \text{ r}4$$

1. $8\overline{)348}$	2. $9\overline{)238}$	3. $8\overline{)747}$	4. $9\overline{)112}$	5. $8\overline{)346}$
6. $7\overline{)314}$	7. $6\overline{)919}$	8. $7\overline{)337}$	9. $6\overline{)419}$	10. $7\overline{)193}$
11. $4\overline{)644}$	12. $5\overline{)519}$	13. $4\overline{)718}$	14. $5\overline{)215}$	15. $4\overline{)790}$

L13 REVIEW A x
3 Digit x 1 Multiplication & Division

$$\begin{array}{r} 504 \\ \underline{\times \quad 3} \\ 1512 \end{array}$$

1. $\begin{array}{r} 254 \\ \underline{\times \quad 3} \end{array}$	2. $\begin{array}{r} 356 \\ \underline{\times \quad 2} \end{array}$	3. $\begin{array}{r} 125 \\ \underline{\times \quad 5} \end{array}$	4. $\begin{array}{r} 500 \\ \underline{\times \quad 4} \end{array}$	5. $\begin{array}{r} 500 \\ \underline{\times \quad 4} \end{array}$
6. $\begin{array}{r} 955 \\ \underline{\times \quad 9} \end{array}$	7. $\begin{array}{r} 726 \\ \underline{\times \quad 8} \end{array}$	8. $\begin{array}{r} 555 \\ \underline{\times \quad 5} \end{array}$	9. $\begin{array}{r} 899 \\ \underline{\times \quad 7} \end{array}$	10. $\begin{array}{r} 936 \\ \underline{\times \quad 7} \end{array}$
11. $\begin{array}{r} 321 \\ \underline{\times \quad 4} \end{array}$	12. $\begin{array}{r} 150 \\ \underline{\times \quad 8} \end{array}$	13. $\begin{array}{r} 133 \\ \underline{\times \quad 6} \end{array}$	14. $\begin{array}{r} 446 \\ \underline{\times \quad 3} \end{array}$	15. $\begin{array}{r} 535 \\ \underline{\times \quad 2} \end{array}$

L13 REVIEW A ÷
3 Digit x 1 Multiplication & Division

$7\overline{)855}$

1. $8\overline{)998}$	2. $7\overline{)856}$	3. $6\overline{)942}$	4. $9\overline{)993}$	5. $5\overline{)845}$
6. $6\overline{)844}$	7. $4\overline{)456}$	8. $3\overline{)296}$	9. $7\overline{)287}$	10. $5\overline{)651}$
11. $2\overline{)773}$	12. $7\overline{)898}$	13. $3\overline{)641}$	14. $6\overline{)738}$	15. $4\overline{)436}$

3/1 DIGIT SPEED TEST

1. $4\overline{)256}$	2. $\begin{array}{r}330\\ \underline{\times\ \ 4}\end{array}$	3. $\begin{array}{r}926\\ \underline{\times\ \ 6}\end{array}$	4. $4\overline{)732}$	5. $\begin{array}{r}116\\ \underline{\times\ \ 7}\end{array}$
6. $\begin{array}{r}420\\ \underline{\times\ \ 7}\end{array}$	7. $3\overline{)415}$	8. $\begin{array}{r}268\\ \underline{\times\ \ 4}\end{array}$	9. $\begin{array}{r}607\\ \underline{\times\ \ 5}\end{array}$	10. $5\overline{)404}$
11. $6\overline{)408}$	12. $\begin{array}{r}500\\ \underline{\times\ \ 3}\end{array}$	13. $\begin{array}{r}706\\ \underline{\times\ \ 0}\end{array}$	14. $\begin{array}{r}604\\ \underline{\times\ \ 9}\end{array}$	15. $5\overline{)400}$
16. $\begin{array}{r}413\\ \underline{\times\ \ 2}\end{array}$	17. $\begin{array}{r}607\\ \underline{\times\ \ 4}\end{array}$	18. $7\overline{)308}$	19. $\begin{array}{r}105\\ \underline{\times\ \ 3}\end{array}$	20. $\begin{array}{r}907\\ \underline{\times\ \ 9}\end{array}$

L14 x
4 Digit x 1 Multiplication & Division

$$\begin{array}{r} 1234 \\ \underline{\times \quad 1} \\ 1234 \end{array}$$

1. $\begin{array}{r} 1356 \\ \underline{\times \quad 7} \end{array}$	2. $\begin{array}{r} 9233 \\ \underline{\times \quad 5} \end{array}$	3. $\begin{array}{r} 4550 \\ \underline{\times \quad 2} \end{array}$	4. $\begin{array}{r} 3500 \\ \underline{\times \quad 2} \end{array}$	5. $\begin{array}{r} 1000 \\ \underline{\times \quad 5} \end{array}$
6. $\begin{array}{r} 8394 \\ \underline{\times \quad 3} \end{array}$	7. $\begin{array}{r} 6428 \\ \underline{\times \quad 4} \end{array}$	8. $\begin{array}{r} 2346 \\ \underline{\times \quad 7} \end{array}$	9. $\begin{array}{r} 5036 \\ \underline{\times \quad 9} \end{array}$	10. $\begin{array}{r} 7309 \\ \underline{\times \quad 1} \end{array}$
11. $\begin{array}{r} 9377 \\ \underline{\times \quad 7} \end{array}$	12. $\begin{array}{r} 6288 \\ \underline{\times \quad 6} \end{array}$	13. $\begin{array}{r} 4273 \\ \underline{\times \quad 9} \end{array}$	14. $\begin{array}{r} 5141 \\ \underline{\times \quad 7} \end{array}$	15. $\begin{array}{r} 7436 \\ \underline{\times \quad 2} \end{array}$

L14 ÷
4 Digit x 1 Multiplication & Division

$$\begin{array}{r} 1234 \\ 1\overline{)1234} \\ \underline{-1234} \end{array}$$

1. $6\overline{)6684}$	2. $4\overline{)2683}$	3. $2\overline{)8697}$	4. $8\overline{)3345}$	5. $9\overline{)1488}$
6. $1\overline{)5000}$	7. $3\overline{)6623}$	8. $5\overline{)7215}$	9. $3\overline{)1256}$	10. $4\overline{)1009}$
11. $5\overline{)2500}$	12. $3\overline{)4178}$	13. $8\overline{)9999}$	14. $2\overline{)1258}$	15. $7\overline{)1502}$

L14 REVIEW A x

4 Digit x 1 Multiplication & Division

$$\begin{array}{r} 3504 \\ \times \quad 5 \\ \hline 17{,}520 \end{array}$$

1. $\begin{array}{r} 3344 \\ \times \quad 5 \\ \hline \end{array}$	2. $\begin{array}{r} 7536 \\ \times \quad 4 \\ \hline \end{array}$	3. $\begin{array}{r} 4550 \\ \times \quad 2 \\ \hline \end{array}$	4. $\begin{array}{r} 8256 \\ \times \quad 6 \\ \hline \end{array}$	5. $\begin{array}{r} 7357 \\ \times \quad 2 \\ \hline \end{array}$
6. $\begin{array}{r} 1256 \\ \times \quad 3 \\ \hline \end{array}$	7. $\begin{array}{r} 8992 \\ \times \quad 4 \\ \hline \end{array}$	8. $\begin{array}{r} 2333 \\ \times \quad 2 \\ \hline \end{array}$	9. $\begin{array}{r} 8356 \\ \times \quad 8 \\ \hline \end{array}$	10. $\begin{array}{r} 1999 \\ \times \quad 3 \\ \hline \end{array}$
11. $\begin{array}{r} 5526 \\ \times \quad 7 \\ \hline \end{array}$	12. $\begin{array}{r} 4056 \\ \times \quad 9 \\ \hline \end{array}$	13. $\begin{array}{r} 6728 \\ \times \quad 5 \\ \hline \end{array}$	14. $\begin{array}{r} 2114 \\ \times \quad 8 \\ \hline \end{array}$	15. $\begin{array}{r} 2000 \\ \times \quad 6 \\ \hline \end{array}$

L14 REVIEW A ÷
4 Digit x 1 Multiplication & Division

816 r7
4) 3267

1. 4) 8873	2. 7) 4569	3. 8) 6995	4. 4) 5573	5. 3) 9905
6. 9) 6458	7. 6) 4753	8. 7) 8080	9. 5) 3655	10. 4) 7866
11. 8) 1119	12. 7) 4568	13. 6) 7896	14. 9) 5544	15. 8) 9987

L14 REVIEW B x
4 Digit x 1 Multiplication & Division

$$\begin{array}{r} 9104 \\ \underline{\times \quad 3} \\ 27.312 \end{array}$$

1. 2346 x 5	2. 9999 x 2	3. 7634 x 3	4. 5101 x 6	5. 2314 x 4
6. 8816 x 9	7. 8721 x 4	8. 1863 x 7	9. 9339 x 2	10. 9884 x 8
11. 8785 x 1	12. 6025 x 3	13. 1080 x 4	14. 6530 x 6	15. 3969 x 5

L14 REVIEW B ÷
4 Digit x 1 Multiplication & Division

866r2
3)2600

1. 8) 4856	2. 7) 6682	3. 5) 8840	4. 9) 7788	5. 4) 7788
6. 3) 1788	7. 2) 6500	8. 3) 8983	9. 6) 5658	10. 7) 1214
11. 9) 3536	12. 4) 1844	13. 7) 2944	14. 3) 1739	15. 8) 4544

4/1 DIGIT SPEED TEST

1. $4\overline{)2056}$	2. 3030 x 3	3. 9026 x 8	4. $4\overline{)7302}$	5. 1016 x 5
6. 4020 x 2	7. $3\overline{)415}$	8. 2608 x 6	9. 6007 x 7	10. $5\overline{)4004}$
11. $6\overline{)4008}$	12. 5000 x 3	13. 7006 x 1	14. 6004 x 2	15. $5\overline{)4000}$
16. 4103 x 8	17. 6007 x 7	18. $7\overline{)3008}$	19. 1005 x 3	20. 9007 x 5

L15 x
2 Digit x 2 Multiplication & Division

$$\begin{array}{r} 12 \\ \underline{x\ \ 12} \\ 144 \\ \underline{+\ 1440} \\ 1584 \end{array}$$

1. 72 x 32	2. 91 x 56	3. 81 x 52	4. 48 x 79	5. 25 x 66
6. 42 x 26	7. 84 x 17	8. 32 x 50	9. 82 x 39	10. 36 x 10
11. 38 x 12	12. 96 x 67	13. 59 x 48	14. 64 x 68	15. 44 x 11

L15 ÷
2 Digit x 2 Multiplication & Division

$$\begin{array}{r} 5 \\ 10\overline{)50} \\ \underline{-50} \end{array}$$

1. $12\overline{)36}$	2. $11\overline{)66}$	3. $17\overline{)35}$	4. $16\overline{)42}$	5. $9\overline{)94}$
6. $12\overline{)14}$	7. $20\overline{)60}$	8. $22\overline{)88}$	9. $13\overline{)63}$	10. $26\overline{)76}$
11. $42\overline{)74}$	12. $30\overline{)40}$	13. $10\overline{)28}$	14. $15\overline{)37}$	15. $16\overline{)38}$

L15 REVIEW A x
2 Digit x 2 Multiplication & Division

$$\begin{array}{r} 58 \\ \underline{\times\ 35} \end{array}$$

1. 49 x 27	2. 73 x 41	3. 11 x 50	4. 40 x 24	5. 34 x 70
6. 15 x 42	7. 92 x 87	8. 23 x 23	9. 21 x 99	10. 39 x 12
11. 82 x 64	12. 88 x 62	13. 27 x 34	14. 86 x 33	15. 40 x 40

L15 REVIEW B x
2 Digit x 2 Multiplication & Division

$$\begin{array}{r} 95 \\ \underline{\times\ 42} \\ 190 \\ \underline{+3800} \\ 3990 \end{array}$$

1. 60 x 82	2. 34 x 29	3. 83 x 41	4. 98 x 50	5. 30 x 25
6. 17 x 55	7. 24 x 33	8. 59 x 58	9. 15 x 68	10. 64 x 34
11. 35 x 50	12. 14 x 38	13. 54 x 70	14. 87 x 96	15. 57 x 10

L15 REVIEW A ÷
2 Digit x 2 Multiplication & Division

$$22\overline{)44}\quad 2$$

1. 27) 47	2. 35) 62	3. 29) 48	4. 24) 44	5. 19) 99
6. 33) 76	7. 44) 98	8. 33) 77	9. 25) 75	10. 14) 93
11. 56) 84	12. 29) 33	13. 76) 98	14. 34) 69	15. 26) 53

L15 REVIEW B ÷
2 Digit x 2 Multiplication & Division

$$12\overline{)12}^{\,1}$$

1. $38\overline{)92}$	2. $46\overline{)73}$	3. $51\overline{)78}$	4. $19\overline{)37}$	5. $78\overline{)88}$
6. $15\overline{)93}$	7. $67\overline{)89}$	8. $48\overline{)62}$	9. $51\overline{)83}$	10. $17\overline{)77}$
11. $47\overline{)52}$	12. $25\overline{)47}$	13. $17\overline{)34}$	14. $69\overline{)76}$	15. $10\overline{)36}$

2/2 DIGIT SPEED TEST

1. $10\overline{)56}$	2. 30×23	3. 26×28	4. $12\overline{)42}$	5. 16×25
6. 70×38	7. $13\overline{)41}$	8. 88×25	9. 97×44	10. $50\overline{)80}$
11. $12\overline{)98}$	12. 70×20	13. 46×52	14. 84×33	15. $30\overline{)75}$
16. 63×78	17. 97×97	18. $20\overline{)78}$	19. 95×63	20. 38×17

16 x
3 Digit x 2 Multiplication & Division

$$\begin{array}{r} 123 \\ \underline{x\ \ 11} \\ 123 \\ \underline{+\ 123} \\ 1356 \end{array}$$

1. 250 x 46	2. 337 x 83	3. 567 x 17	4. 133 x 63	5. 129 x 34
6. 182 x 21	7. 861 x 24	8. 356 x 86	9. 167 x 18	10. 125 x 55
11. 136 x 54	12. 546 x 38	13. 250 x 20	14. 217 x 88	15. 151 x 12

L16 ÷
3 Digit x 2 Multiplication & Division

1234
1)1234
-1234

1. 334 x 65	2. 756 x 54	3. 470 x 32	4. 826 x 46	5. 357 x 62
6. 256 x 8 3	7. 992 x 7 7	8. 333 x 62	9. 856 x 48	10. 999 x 73
11. 526 x 87	12. 596 x 29	13. 728 x 85	14. 184 x 98	15. 207 x 46

L16 REVIEW A x
3 Digit x 2 Multiplication & Division

$$\begin{array}{r} 2\,95 \\ \underline{\times\ \ 33} \end{array}$$

1. 363 x 19	2. 476 x 13	3. 582 x 82	4. 141 x 14	5. 683 x 94
6. 756 x 13	7. 674 x 76	8. 459 x 33	9. 760 x 90	10. 867 x 21
11. 240 x 26	12. 514 x 37	13. 191 x 91	14. 758 x 26	15. 334 x 19

L16 REVIEW B x
3 Digit x 2 Multiplication & Division

$$\begin{array}{r} 945 \\ \times\ 16 \\ \hline \end{array}$$

1. 141 x 10	2. 610 x 10	3. 963 x 10	4. 584 x 10	5. 723 x 10
6. 644 x 12	7. 789 x 12	8. 848 x 12	9. 126 x 12	10. 444 x 12
11. 846 x 44	12. 623 x 70	13. 184 x 33	14. 397 x 47	15. 583 x 20

L16 REVIEW A ÷
3 Digit x 2 Multiplication & Division

$13\overline{)250}$

1. $44\overline{)992}$	2. $56\overline{)996}$	3. $64\overline{)872}$	4. $88\overline{)670}$	5. $38\overline{)546}$
6. $25\overline{)277}$	7. $42\overline{)414}$	8. $73\overline{)409}$	9. $91\overline{)663}$	10. $47\overline{)331}$
11. $68\overline{)771}$	12. $33\overline{)705}$	13. $18\overline{)296}$	14. $22\overline{)888}$	15. $30\overline{)701}$

L16 REVIEW B ÷
3 Digit x 2 Multiplication & Division

21)870

1. 16)555	2. 27)888	3. 33)333	4. 19)444	5. 25)250
6. 13)999	7. 15)553	8. 67)333	9. 43)681	10. 38)222
11. 29)222	12. 47)777	13. 36)443	14. 66)991	15. 40)900

3/2 DIGIT SPEED TEST

1. 10) 506	2. 300 x 33	3. 260 x 68	4. 12) 402	5. 160 x 65
6. 700 x 38	7. 13) 410	8. 880 x 25	9. 970 x 34	10. 50) 800
11. 12) 980	12. 700 x 60	13. 460 x 62	14. 840 x 63	15. 30) 705
16. 630 x 78	17. 970 x 37	18. 20) 708	19. 950 x 33	20. 380 x 37

L17 x
4 Digit x 2 Multiplication & Division

$$\begin{array}{r} 1234 \\ \underline{\times \quad 11} \\ 1234 \\ \underline{+\ 1234} \\ 13{,}574 \end{array}$$

1. 3344 x 25	2. 7536 x 54	3. 4550 x 22	4. 8256 x 14	5. 7357 x 42
6. 1256 x 13	7. 8992 x 34	8. 2333 x 12	9. 8356 x 38	10. 1999 x 23
11. 5526 x 27	12. 4056 x 29	13. 6728 x 15	14. 2114 x 18	15. 2000 x 46

L17 ÷

4 Digit x 2 Multiplication & Division

1202 R10
12)1234

1. 48)9641	2. 54)7214	3. 33)6712	4. 26)3425	5. 21)4945
6. 44)6477	7. 14)6003	8. 45)7787	9. 44)1202	10. 44)6302
11. 85)9212	12. 17)2784	13. 15)6203	14. 14)8903	15. 16)1488

L17 REVIEW A x
4 Digit x 2 Multiplication & Division

$22\overline{)4624}$

1. 2000 x 15	2. 5000 x 34	3. 6800 x 77	4. 8200 x 21	5. 6400 x 50
6. 3100 x 24	7. 1600 x 16	8. 2222 x 22	9. 1400 x 17	10. 2500 x 25
11. 5583 x 13	12. 4451 x 13	13. 4400 x 10	14. 6500 x 20	15. 1000 x 10

L17 REVIEW B x
4 Digit x 2 Multiplication & Division

$$\begin{array}{r} 1234 \\ \times \quad 11 \\ \hline \end{array}$$

1. 3344 x 65	2. 7536 x 84	3. 4550 x 42	4. 8256 x 76	5. 7357 x 92
6. 1256 x 83	7. 8992 x 64	8. 2333 x 82	9. 8356 x 78	10. 1999 x 93
11. 5526 x 97	12. 4056 x 69	13. 6728 x 55	14. 2114 x 68	15. 2000 x 96

L17 REVIEW A ÷
4 Digit x 2 Multiplication & Division

$26\overline{)2677}$

1. $48\overline{)9641}$	2. $54\overline{)7214}$	3. $33\overline{)6712}$	4. $26\overline{)3425}$	5. $21\overline{)4945}$
6. $44\overline{)6477}$	7. $14\overline{)6003}$	8. $45\overline{)7787}$	9. $44\overline{)1202}$	10. $44\overline{)6302}$
11. $85\overline{)9212}$	12. $17\overline{)2784}$	13. $15\overline{)6203}$	14. $14\overline{)8903}$	15. $16\overline{)1488}$

L17 REVIEW B ÷

4 Digit x 2 Multiplication & Division

$12\overline{)2424}$

1. $27\overline{)3336}$	2. $15\overline{)3738}$	3. $28\overline{)5888}$	4. $34\overline{)7070}$	5. $14\overline{)2914}$
6. $12\overline{)3521}$	7. $10\overline{)7112}$	8. $78\overline{)8286}$	9. $31\overline{)1705}$	10. $83\overline{)7311}$
11. $46\overline{)3351}$	12. $71\overline{)3151}$	13. $81\overline{)1300}$	14. $23\overline{)1538}$	15. $37\overline{)8299}$

4/2 DIGIT SPEED TEST

1. $12\overline{)1006}$	2. 1000×23	3. 3060×48	4. $22\overline{)5002}$	5. 7060×25
6. 5000×38	7. $13\overline{)3010}$	8. 4080×35	9. 2070×54	10. $30\overline{)1000}$
11. $12\overline{)4080}$	12. 8000×70	13. 5060×92	14. 3040×93	15. $30\overline{)5005}$
16. 2030×28	17. 3070×57	18. $20\overline{)4008}$	19. 7050×43	20. 8080×97

L18 x
3 Digit x 3 Multiplication & Division

$$\begin{array}{r} 123 \\ \times\ 111 \\ \hline \end{array}$$

1. 344 x 45	2. 536 x 88	3. 550 x 32	4. 256 x 86	5. 357 x 62
6. 256 x 83	7. 992 x 94	8. 333 x 92	9. 356 x 38	10. 999 x 43
11. 526 x 97	12. 596 x 39	13. 728 x 55	14. 114 x 78	15. 280 x 56

L18 ÷

3 Digit x 3 Multiplication & Division

$11\overline{)123}$

1. $120\overline{)400}$	2. $130\overline{)960}$	3. $125\overline{)500}$	4. $400\overline{)620}$	5. $210\overline{)801}$
6. $240\overline{)450}$	7. $220\overline{)480}$	8. $180\overline{)622}$	9. $220\overline{)440}$	10. $120\overline{)190}$
11. $135\overline{)500}$	12. $150\overline{)300}$	13. $100\overline{)180}$	14. $170\overline{)702}$	15. $201\overline{)350}$

L18 REVIEW A x
3 Digit x 3 Multiplication & Division

$$\begin{array}{r} 123 \\ \underline{\times\ 111} \end{array}$$

1. $\begin{array}{r} 617 \\ \underline{\times\ 340} \end{array}$	2. $\begin{array}{r} 726 \\ \underline{\times\ 184} \end{array}$	3. $\begin{array}{r} 256 \\ \underline{\times\ 651} \end{array}$	4. $\begin{array}{r} 440 \\ \underline{\times\ 443} \end{array}$	5. $\begin{array}{r} 627 \\ \underline{\times\ 187} \end{array}$
6. $\begin{array}{r} 423 \\ \underline{\times\ 324} \end{array}$	7. $\begin{array}{r} 876 \\ \underline{\times\ 917} \end{array}$	8. $\begin{array}{r} 551 \\ \underline{\times\ 151} \end{array}$	9. $\begin{array}{r} 227 \\ \underline{\times\ 446} \end{array}$	10. $\begin{array}{r} 123 \\ \underline{\times\ 123} \end{array}$
11. $\begin{array}{r} 250 \\ \underline{\times\ 480} \end{array}$	12. $\begin{array}{r} 403 \\ \underline{\times\ 906} \end{array}$	13. $\begin{array}{r} 918 \\ \underline{\times\ 926} \end{array}$	14. $\begin{array}{r} 544 \\ \underline{\times\ 870} \end{array}$	15. $\begin{array}{r} 238 \\ \underline{\times\ 146} \end{array}$

L18 REVIEW A ÷
3 Digit x 3 Multiplication & Division

$252\overline{)4564}$

1. $444\overline{)792}$	2. $330\overline{)760}$	3. $222\overline{)550}$	4. $412\overline{)793}$	5. $338\overline{)876}$
6. $216\overline{)904}$	7. $446\overline{)583}$	8. $154\overline{)711}$	9. $308\overline{)415}$	10. $102\overline{)708}$
11. $278\overline{)632}$	12. $415\overline{)834}$	13. $124\overline{)641}$	14. $804\overline{)906}$	15. $115\overline{)706}$

L18 REVIEW B x
3 Digit x 3 Multiplication & Division

$$\begin{array}{r} 234 \\ \times\ 234 \\ \hline \end{array}$$

1. 223 x 142	2. 387 x 454	3. 617 x 907	4. 515 x 634	5. 107 x 223
6. 544 x 407	7. 931 x 658	8. 741 x 800	9. 414 x 276	10. 719 x 917
11. 362 x 454	12. 109 x 756	13. 673 x 586	14. 229 x 338	15. 200 x 786

L18 REVIEW B ÷

3 Digit x 3 Multiplication & Division

$12\overline{)2424}$

1. $334\overline{)670}$	2. $240\overline{)683}$	3. $440\overline{)880}$	4. $304\overline{)608}$	5. $501\overline{)944}$
6. $333\overline{)704}$	7. $119\overline{)670}$	8. $209\overline{)614}$	9. $220\overline{)440}$	10. $403\overline{)377}$
11. $211\overline{)704}$	12. $709\overline{)994}$	13. $124\overline{)641}$	14. $414\overline{)814}$	15. $319\overline{)699}$

3/3 DIGIT SPEED TEST

1. $12\overline{)1006}$	2. 1000×23	3. 3060×48	4. $22\overline{)5002}$	5. 7060×25
6. 5000×38	7. $13\overline{)3010}$	8. 4080×35	9. 2070×54	10. $30\overline{)1000}$
11. $12\overline{)4080}$	12. 8000×70	13. 5060×92	14. 3040×93	15. $30\overline{)5005}$
16. 2030×28	17. 3070×57	18. $20\overline{)4008}$	19. 7050×43	20. 8080×97

L19 x
4 Digit x 3 Multiplication & Division

```
       1234
  x     111
      1234
     1234
    1234. ...
  136,974
```

1. 5683 x 191	2. 4706 x 336	3. 5374 x 416	4. 2613 x 458	5. 1907 x 283
6. 2452 x 817	7. 2673 x 334	8. 7861 x 258	9. 7785 x 434	10. 1545 x 337
11. 2672 x 485	12. 2672 x 338	13. 4256 x 276	14. 1888 x 976	15. 8888 x 8888

L19 ÷
4 Digit x 3 Multiplication & Division

111)1234

1. 405)3450	2. 290)1476	3. 405)1050	4. 703)1248	5. 242)8306
6. 193)9870	7. 159)6570	8. 204)4899	9. 301)4456	10. 216)7899
11. 302)6088	12. 403)4876	13. 111)5118	14. 801)7688	15. 156)8401

19 REVIEW A x
4 Digit x 3 Multiplication & Division

$$\begin{array}{r} 4530 \\ \times \quad 227 \\ \hline \end{array}$$

1. $\begin{array}{r} 2624 \\ \times \quad 408 \\ \hline \end{array}$	2. $\begin{array}{r} 2725 \\ \times \quad 418 \\ \hline \end{array}$	3. $\begin{array}{r} 2826 \\ \times \quad 428 \\ \hline \end{array}$	4. $\begin{array}{r} 2927 \\ \times \quad 438 \\ \hline \end{array}$	5. $\begin{array}{r} 3276 \\ \times \quad 517 \\ \hline \end{array}$
6. $\begin{array}{r} 3377 \\ \times \quad 527 \\ \hline \end{array}$	7. $\begin{array}{r} 3478 \\ \times \quad 537 \\ \hline \end{array}$	8. $\begin{array}{r} 3579 \\ \times \quad 547 \\ \hline \end{array}$	9. $\begin{array}{r} 3683 \\ \times \quad 559 \\ \hline \end{array}$	10. $\begin{array}{r} 3786 \\ \times \quad 563 \\ \hline \end{array}$
11. $\begin{array}{r} 4372 \\ \times \quad 628 \\ \hline \end{array}$	12. $\begin{array}{r} 4473 \\ \times \quad 638 \\ \hline \end{array}$	13. $\begin{array}{r} 4474 \\ \times \quad 649 \\ \hline \end{array}$	14. $\begin{array}{r} 4575 \\ \times \quad 653 \\ \hline \end{array}$	15. $\begin{array}{r} 9999 \\ \times \quad 9999 \\ \hline \end{array}$

L19 REVIEW A ÷
4 Digit x 3 Multiplication & Division

12) 2424

1. 215) 3450	**2.** 307) 5314	**3.** 296) 7801	**4.** 190) 8702	**5.** 130) 5477
6. 260) 4078	**7.** 451) 9211	**8.** 570) 8989	**9.** 520) 7840	**10.** 330) 8950
11. 270) 2886	**12.** 150) 3500	**13.** 216) 2166	**14.** 709) 9077	**15.** 205) 2055

L19 REVIEW B x
4 Digit x 3 Multiplication & Division

$$\begin{array}{r} 1234 \\ \times \underline{\quad 111} \end{array}$$

1. 4883 x 622	2. 4993 x 633	3. 4773 x 522	4. 4663 x 433	5. 4596 x 337
6. 4486 x 222	7. 4385 x 386	8. 5873 x 634	9. 6783 x 792	10. 7988 x 324
11. 7986 x 437	12. 7868 x 734	13. 7995 x 487	14. 8647 x 395	15. 6666 x 6666

L19 REVIEW B ÷

4 Digit x 3 Multiplication & Division

300) 4000

1. 355) 4884	**2.** 205) 6700	**3.** 296) 8745	**4.** 405) 7605	**5.** 305) 4780
6. 708) 6992	**7.** 333) 6988	**8.** 415) 7613	**9.** 602) 8000	**10.** 503) 5030
11. 401) 4010	**12.** 206) 6755	**13.** 703) 8300	**14.** 916) 9980	**15.** 701) 7015

4/3 DIGIT SPEED TEST

1. $402\overline{)1006}$	2. 1000 x 253	3. 3060 x 428	4 $220\overline{)5002}$	5. 7060 x 200
6. 5000 x 300	7. $100\overline{)3010}$	8. 4080 x 300	9. 2070 x 500	10. $200\overline{)1000}$
11. $100\overline{)4080}$	12. 8000 x 400	13. 5060 x 900	14. 3040 x 900	15. $300\overline{)5005}$
16. 9030 x 400	17. 8070 x 500	18. $200\overline{)7008}$	19. 7050 x 400	20. 8080 x 900

L20 x
4 Digit x 4 Multiplication & Division

$$\begin{array}{r} 4100 \\ \times\ 3027 \\ \hline \end{array}$$

1. 1274 x 1333	2. 4708 x 9961	3. 1414 x 3236	4. 1981 x 2607	5. 2207 x 3439
6. 2448 x 1992	7. 3738 x 2434	8. 7787 x 7229	9. 9920 x 2276	10. 5343 x 1382
11. 1707 x 4486	12. 4409 x 9910	13. 5673 x 2685	14. 4408 x 4747	15. 3333 x 3333

L20 ÷
4 Digit x 4 Multiplication & Division

$$1234 \overline{)1234}^{\,1234}$$

1. $2110\overline{)3450}$	2. $1604\overline{)3605}$	3. $1324\overline{)3688}$	4. $2155\overline{)7684}$	5. $1550\overline{)6441}$
6. $2264\overline{)8440}$	7. $2300\overline{)6604}$	8. $1848\overline{)8840}$	9. $3406\overline{)9412}$	10. $1460\overline{)2830}$
11. $2602\overline{)4848}$	12. $1701\overline{)8301}$	13. $1212\overline{)2456}$	14. $1603\overline{)8843}$	15. $1260\overline{)3684}$

L20 REVIEW A x
4 Digit x 4 Multiplication & Division

$$\begin{array}{r} 6100 \\ \times\ 2245 \\ \hline \end{array}$$

1. 1448 x 1967	2. 1659 x 2783	3. 2391 x 1115	4. 2267 x 1467	5. 1783 x 1986
6. 3424 x 1977	7. 6217 x 9448	8. 5563 x 5471	9. 2276 x 2557	10. 3376 x 2448
11. 7266 x 3563	12. 7946 x 3345	13. 2784 x 1405	14. 7636 x 9646	15. 7777 x 7777

L20 REVIEW B x
4 Digit x 4 Multiplication & Division

$$\begin{array}{r} 5500 \\ \times\ 3300 \\ \hline \end{array}$$

1. 7848 x 1926	2. 4559 x 2737	3. 7791 x 1155	4. 3867 x 1499	5. 4877 x 6234
6. 4509 x 2246	7. 7737 x 9287	8. 3990 x 9485	9. 2626 x 1447	10. 9877 x 3245
11. 4877 x 9607	12. 2688 x 1525	13. 4984 x 3434	14. 2097 x 3458	15. 2872 x 2654

L20 REVIEW A ÷
4 Digit x 4 Multiplication & Division

2000)‾4000

1. 1804)‾3604	2. 1612)‾3622	3. 2204)‾4408	4. 1214)‾3804	5. 2601)‾3815
6. 3301)‾7515	7. 4203)‾8406	8. 2515)‾5030	9. 1701)‾4501	10. 2216)‾4305
11. 6000)‾8200	12. 1455)‾3688	13. 2400)‾4807	14. 1355)‾2700	15. 2414)‾4848

L20 REVIEW B ÷
4 Digit x 4 Multiplication & Division

1000)5000

1. 3030)6065	2. 4404)8808	3. 5301)9554	4. 4200)8400	5. 2712)5451
6. 1680)4500	7. 3348)7655	8. 4604)6299	9. 1240)4880	10. 1804)3668
11. 1454)7640	12. 2140)4870	13. 3200)6412	14. 1650)7700	15. 2414)4828

4/4 DIGIT SPEED TEST

1. 6000) 8200	2. 1000 x 2443	3. 3060 x 8828	4. 1355) 2700	5. 7060 x 5500
6. 5000 x 7850	7. 1454) 7640	8. 4080 x 8740	9. 2070 x 7440	10. 3200) 6412
11. 4604) 6299	12. 8000 x 4747	13. 5060 x 5550	14. 3040 x 3250	15. 1612) 3622
16. 9030 x 4630	17. 8070 x 5800	**10.** 2216) 4305	19. 7050 x 8820	20. 8080 x 5280

WHAT'S COMING NEXT; MATHEMATICALLY SPEAKING

Up next on your mathematical journey is what I like to call the essentials. The essentials comprise all the mathematical skills you need to master on your k-12 journey. After you master the essentials, you should move on to Algebra 1.

Since you have mastered the operations, you will find the essentials to be much easier and far more refreshing than jumping into them without having mastered the skills in this book. The Finding Success in Mathematics program is specially formatted to make you into a mathematical genius ready for every mathematical Olympiad that awaits you.

If you have enjoyed learning Algebra 1 and would like more practice or a more in-depth study, please let your instructor know. If you are in grades 9-12, you might want to take a course at the local community college. You might also want to offer your services as a tutor for other high school students who would like more help understanding Algebra 1.

www.ingramcontent.com/pod-product-compliance
Lightning Source LLC
LaVergne TN
LVHW070951210926
848482LV00002B/176